Springer Series in
Computational
Mathematics

11

Jean-Paul Delahaye

Sequence Transformations

With an Introduction by
Claude Brezinski

Springer-Verlag
Berlin Heidelberg New York
London Paris Tokyo

Jean-Paul Delahaye
Laboratoire d'Informatique Fondamentale de Lille
Université des Sciences et Techniques de Lille Flandres Artois
U.F.R. d'I.E.E.A., Bât. M3, F-59655 Villeneuve d'Ascq Cedex,
France

With 164 Figures

Mathematics Subject Classification (1980): 65 B 05, 65 B 99,
68 C 05, 68 C 40

ISBN-13:978-3-642-64802-1   e-ISBN-13:978-3-642-61347-0
DOI: 10.1007/978-3-642-61347-0

Library of Congress Cataloging-in-Publication Data
Delahaye, Jean-Paul. Sequence transformations.
(Springer series in computational mathematics ; 11)
Includes bibliographies and index.
1. Sequences (Mathematics) 2. Transformations (Mathematics) 3. Numerical
analysis—Acceleration of convergence. I. Title. II. Series.
QA292.D45 1988 515'.24 88-4527
ISBN-13:978-3-642-64802-1(U.S.)

2145/3140-543210

CONTENTS

Chapter 3
Algorithms for Extracting Convergent Subsequences   93

Chapter 4
The Partially Ordered Systems of Accelerable Families   125

Chapter 5
Non-Accelerable Families of Sequences   151

INTRODUCTION

Claude BREZINSKI

It is my great pleasure to write this introduction because, as will become obvious to the reader after the first few pages, this work is a major contribution to the theory and practice of convergence acceleration methods. The impact of Jean-Paul DELAHAYE's results on the subject would be much better appreciated if one had some knowledge of the past efforts in the field. Thus, my aim is to give a brief history of convergence acceleration methods.

The first methods derived and used were linear summation processes. The sequence (S_n) to be accelerated was transformed into a sequence (T_n), where :

$$T_n = \sum_{i=0}^{\infty} a_{ni} S_i \qquad n = 0, 1, \ldots,$$

the a_{ni}'s being independant of the initial sequence. A summation process is completely determined by the matrix $A = (a_{ni})$. In practice, only a finite number of a_{ni}'s are non-zero for all n. Among such processes are those named after EULER, CESARO and HOLDER. The convergence of the sequence (T_n) is completely settled by the well-known Toeplitz's theorem on the matrix A.

In some particular cases, summation processes can accelerate the convergence but non-linear sequence transformation are usually better. This is the reason that numerical analysts soon devoted their efforts to non-linear processes.

The most popular non-linear acceleration algorithm is probably the so-called Δ^2 process attributed to A.C. AITKEN (1926). It consists of transforming (S_n) into (T_n), where

$$T_n = S_n - (\Delta S_n)^2 / \Delta^2 S_n \qquad n = 0, 1, \ldots \qquad (1)$$

The Aitken's aim when he proposed this method, which he called δ^2, was to accelerate the convergence of Bernouilli's method for computing the dominant zero of a polynomial. Aitken pointed out that the same method was incidentally obtained by H. VON NAEGELSBACH in 1876 while studying Furstenau's method for solving equations.

The process was also given by J.C. MAXWELL in his treatise on electricity in 1892. But neither Naegelsbach nor Maxwell used this algorithm to accelerate convergence. Maxwell's aim was to find the equilibrium position of a pointer oscillating with an exponentially damped simple harmonic motion from three experimental measurements. If this is expressed in term of sequences, we have :

$$T_n = S \quad \forall \, n$$

for every sequence (S_n) of the form :

$$S_n = S + \alpha \, \gamma^n \quad \forall \, n$$

with $\gamma \neq 1$ or, in other words, such that :

$$a_0(S_n - S) + a_1(S_{n+1} - S) = 0 \quad \forall \, n$$

with $a_0 \, a_1 \neq 0$ and $a_0 + a_1 \neq 0$.

Let me mention that Aitken's process was used in 1674 by SEKI KOWA, who is considered as the greatest Japanese mathematician. Let c_i be the perimeter of the polygon with 2^i sides inscribed in a circle of diameter one. To derive a better approximation S of Π , Seki made use of the formula :

$$S = c_{16} + \frac{(c_{16} - c_{15}) \, (c_{17} - c_{16})}{(c_{16} - c_{15}) - (c_{17} - c_{16})}$$

which is exactly Aitken's process.

The next step was to generalize Aitken's process (1) to the case where :

$$S_n = S + \alpha_1 \, \gamma_1{}^n + \ldots + \alpha_k \, \gamma_k{}^n \quad \forall \, n \quad (2)$$

with $\forall \, i$, $\gamma_i \neq 1$ or, in other words and more generally, to sequences satisfying :

$$a_0(S_n - S) + \ldots + a_k(S_{n+k} - S) = 0 \quad \forall \, n \quad (3)$$

with $a_k \neq 0$ and $a_0 + \ldots + a_k \neq 0$.

Starting with (2) , the case $k = 2$ was studied by Maxwell in his book.

The general case was treated independently by T.H. O'BEIRNE in 1947 from (2) and by D. SHANKS in 1949 from (2) and in 1955 from (3) .

This sequence transformation, now known as Shanks' transformation, consists in transforming (S_n) into a set of sequences $\{(e_k(S_n))\}$ given by

$$e_k(S_n) = \cfrac{\begin{vmatrix} S_n & S_{n+1} & \cdots & S_{n+k} \\ S_{n+k} & S_{n+k+1} & \cdots & S_{n+2k} \\ \hline S_{n+k} & S_{n+k+1} & \cdots & S_{n+2k} \end{vmatrix}}{\begin{vmatrix} \Delta^2 S_n & \cdots & \Delta^2 S_{n+k-1} \\ \hline \Delta^2 S_{n+k-1} & \cdots & \Delta^2 S_{n+2k-2} \end{vmatrix}} \qquad k,n \geq 0 \quad (4)$$

If $k = 1$, $e_1(S_n)$ is identical with Aitken's Δ^2 process. If (S_n) satisfies (3) ((2) is a particular case of (3)) then, by construction of the transformation, $e_k(S_n) = S$, $\forall n$.

The ratio of determinants (4) was also obtained by R.J. SCHMIDT in 1941 while studying a method for solving systems of linear equations, thus anticipating one of the applications of Shanks' transformation. This ratio is implicitly contained in the determinantal formula for Padé approximants given by C.G.J. JACOBI in 1846.

The numerical computation of $e_k(S_n)$ was achived by O'BEIRNE and SHANKS by separately calculating the numerators and denominators; defining the Henkel determinants by

$$H_k(u_n) = \begin{vmatrix} u_n & \cdots & u_{n+k-1} \\ \hline u_{n+k-1} & \cdots & u_{n+2k-2} \end{vmatrix} \quad ,$$

we have the recurrence relationship :

$$H_0(u_n) = 1 \qquad H_1(u_n) = u_n$$

$$H_{k+1}(u_{n-1}) H_{k-1}(u_{n+1}) = H_k(u_{n-1}) H_k(u_{n+1}) - \left[H_k(u_n)\right]^2$$

and thus :

$$e_k(S_n) = H_{k+1}(S_n)/H_k(\Delta^2 S_n) .$$

In 1956, P. WYNN obtained a recursive method, called the ε-algorithm, for computing the $e_k(S_n)$'s without explicitly computing the Hankel determinants appearing in (4). It is as follows :

$$\varepsilon_{-1}(n) = 0 \qquad \varepsilon_0(n) = S_n \qquad n = 0,1$$

$$\varepsilon_{k+1}(n) = \varepsilon_{k-1}(n+1) + 1/(\varepsilon_k(n+1) - \varepsilon_k(n)) \qquad k,n = 0,1,\ldots$$

The fundamental result is that $\forall\ n,k$:

$$\varepsilon_{2k}(n) = e_k(S_n) \ .$$

After the discovery of ε-algorithm and the numerous papers on the subject published by P. WYNN between 1956 and 1970, the study of convergence acceleration methods entered into a more active and fruitful stage.

However, before reviewing these developments, we shall look back to a linear extrapolation method which is also well known and has been widely used; I would like to talk about the so-called Richardson's extrapolation process, whose origins are much older.

The problem of the computation of Π is a very old one. Archimedes solved it by using inscribed and circumscribed polygons to the circle. We have :

$$T'_n = n \sin \frac{\Pi}{n} < \Pi < T_n = n \ tg \ \frac{\Pi}{n}$$

where T'_n and T_n are respectively the perimeters of these polygons with n sides. Taking $n = 96$, Archimedes obtained a value of Π with two exact figures. In 1593, Romanus gave 15 digits with $n = 2^{30}$ and Ludolph VAN CEULEN obtained 35 figures wwith $n = 2^{62}$ in 1610.

If n is replaced by $1/h$ in T_n , we have the expansion :

$$T(h) = \Pi + \frac{\Pi^3}{3} h^2 + \frac{2\Pi^5}{15} h^4 + \ldots = \Pi + a_1 h^2 + a_2 h^4 + \ldots$$

In 1654, C. HUYGENS had the idea of combining the values of $T(h)$ and $T(h/2)$ to get a better result. He showed that :

$$T_1(h) = \frac{1}{3} \left[4\ T(h/2) - T(h) \right] = \Pi - \frac{\Pi^5}{30} h^4 + \frac{\Pi^7}{252} h^6 - \ldots$$

with $h = 2^{-30}$ we obtain Π with 35 exact figures. Of course the same process can be repeated with $T_1(h)$. The idea of linear iterative extrapolation was born. Is was further developed by W.F. SHEPPARD in 1900 and R.M. MILNE in 1903, but its systematic study is due to L.F. RICHARDSON in 1910 and later.

$T_1(h)$ as given above can be obtained by linear extrapolation at the point zero. Of course, it is possible to use extrapolation at zero by polynomials of higher degrees. Thus we obtain a set of sequences $\{(T_k^{(n)}\}$ given by :

$$T_k^{(n)} = \frac{\begin{vmatrix} S_n & \cdots\cdots\cdots & S_{n+k+1} \\ h_n & \cdots\cdots\cdots & h_{n+k+1} \\ \cdots\cdots\cdots\cdots\cdots \\ h^k{}_n & \cdots\cdots & h^k{}_{n+k+1} \end{vmatrix}}{\begin{vmatrix} 1 & \cdots\cdots\cdots & 1 \\ h_n & \cdots\cdots\cdots & h_{n+k+1} \\ \cdots\cdots\cdots\cdots\cdots \\ h^k{}_n & \cdots\cdots & h^k{}_{n+k+1} \end{vmatrix}} \quad , \quad (5)$$

where $S_n = T(h_n)$. The numerical values of the $T_k^{(n)}$'s can be obtained recursively without computing the determinants appearing in (5) :

$$T_0^{(n)} = S_n \quad n = 0,1,\ldots$$

$$T_{k+1}^{(n)} = \frac{T_k^{(n)} \, h_{n+k+1} - T_k^{(n+)} \, h_n}{h_{n+k+1} - h_n} \quad k,n = 0,1,\ldots$$

This algorithm, known as Richardson's extrapolation process, is a direct consequence of Neville-Aitken scheme for the calculation of interpolation polynomials.

If $T(h_n)$ represents the value obtained by the trapezoidal rule with the step h_n for computing a definite integral and if we choose $h_{n+1} = h_n/2$, then Richardson's method is identical with the method heuristically proposed by W. ROMBERG in 1955. The connection between both methods and the study of the convergence of Richardson's process were taken up by J.P. LAURENT in 1964. He gave a necessary and sufficient condition on (h_n) in order that $(T_k^{(n)})$ converges to the same limit as (S_n) for fixed n.

Another process based on extrapolation by rational functions was given by WYNN in 1956. It consists of using reciprocal differences as calculated by the ρ-algorithm.

During this period, many other acceleration methods were proposed without any justification other than their efficiency in some particular cases : the w-process of S. LUBKIN (1952), OVERHOLT's method (1965), G transformation of W.D. CLARK, H.L. GRAY, T.A. ATCHINSON and W.C. PYE (1967), my own Θ-algorithm (1971), GERMAIN-BONNE's transformation (which consists in taking $h_n = \Delta \, S_n$ in Richardson's process (1971)), the various transformations due to D. LEVIN (1973), and extrapolation in the least square sense of F. CORDELLIER (1978).

Some of these algorithms were presented with convergence and acceleration results since, in general, non-linear transformations do not transform every convergent sequence into a sequence converging to the same limit and, a fortiori, do not accelerate the convergence of every sequence. One has to find, or to characterize, the sets of sequences for which the processes are regular and those which are accelerated. Usually such results are quite difficult to obtain due to the non-linearity of the algorithms.

At the same time as these developments, several applications of convergence acceleration methods were proposed. For example, the ε-algorithm can be used for solving systems of non-linear equations, where it provides a quadratic method without calculating any derivative or inverting any matrix (C. BREZINSKI, 1970, E. GEKELER, 1972). This method has been used for solving boundary value problems for ordinary differential equations, for the implementation of semi-implicit Runge-Kutta methods and for the estimation of parameters in mathematical models. The ε-algorithm generalizes Bernouilli's method for the dominant zero of a polynomial and the power method for the dominant eigenvalue of a matrix.

The ε and ρ-algorithms were used in quadrature methods, for the inversion of Laplace transforms and for the construction of A-stable methods for ordinary differential equations.

The implementation of convergence acceleration methods on a computer gives rise to several problems due to the propagation of rounding errors. Usually such algorithms are very numerically instable and particular rules for avoiding instability have to be used. The study of these problems was initiated by P. WYNN in 1959 and followed by F. CORDELLIER, who generalized Wynn's particular rules in 1973.

The experience gained during this period was sufficient to study more general ideas on convergence acceleration methods and to derive more general sequence transformations and algorithms. It was time to throw some light on the properties that a sequence transformation must have in order to accelerate the convergence of certain sets of sequences; The first tentative step in such a formulation was due to R. PENNACHI in 1968. His work was generalized in 1973 by B. GERMAIN-BONNE, who proposed a framework gathering many methods and showed how to construct new algorithms for accelerating certain classes of sequences. In the same context, the optimal summation methods of J. BARANGER (1973) must be mentioned.

In almost all of the preceding research, the approach used was first to construct a sequence transformation and then to study its properties : kernel, convergence and acceleration results, applications, stability,...

The complementary point of view, which consists of choosing a class of sequences and then asking if it can be accelerated and by what algorithm, was explored by B. GERMAIN-BONNE in 1978. This point of view leads to a more profound understanding of convergence acceleration.

The second synthetic approach concerns the algorithms. We saw that many sequence transformations are available and that each of them, usually given as a ratio of determinants, can be implemented by a (different) algorithm. A synthesis of extrapolation processes and a general algorithm gathering most of the algorithms actually known was given by T. HAVIE in 1979 and by myself in 1980. The complete theory was very recently built by G. MUHLBACH.

The work of Jean-Paul DELAHAYE brings new and very interesting and powerful ideas on convergence acceleration. First, he precisely defineswhat a sequence algorithmic transformation is. Then he studies classes of sequences and gives a negative property : the remanence. If a set of sequences has this property, then it is not accelerable by a single algorithm. This is a very interesting result, showing the possibilities and the limitations of convergence acceleration methods.

This notion, developed in collaboration with B. GERMAIN-BONNE, is widely studied in the book. In my opinion, another very interesting concept introduced by DELAHAYE is that of automatic selection procedures between sequence transformations. When one has to accelerate a given sequence, it is not often easy to choose an algorithm. Thus the idea is to use simultaneously several methods and, at each step, to select one of the answers according to some criterion.

Of course the book contains many other ideas and results. One can say that our point of view on convergence acceleration has been widely enlarged and changed by the work of Jean-Paul DELAHAYE.

P R E S E N T A T I O N

The notion of sequence transformation as used in numerical analysis (especially in convergence acceleration) has been formalised several times ([1],[3],[5],[6]).

However, these formalisations are only partial and far from include all the transformations really used. On the other hand, the notion of a sequence transformation as it should be defined in the theory of recursivity involves technically difficult aspects (Turing machines, calculable reals) and is not in the scope and in the form in which numerical analysts solve their problems.

The consequence of this situation is that it is not possible to state and prove results that appear natural and certainly true about the intrinsic limitations of the sequence transformations one can use in numerical analysis. Another drawback is the non-existence of general methods for classifying sequence transformations.

In optimization, the theory of general algorithms ([2],[4],[7]) (a special algorithmic theory not related to the theory of recursivity) allows us to formalize and to synthesize almost all the known optimization methods and gives a better understanding of the problems. Analogously, we would like to develop a general algorithmic theory on sequence transformations.

Thus the most important part of our work is an attempt to formalize, to study, to develop and to use a theory of sequence transformations in numerical analysis. This theory allows to obtain limitation results and classification results. We think that our work shows the value of such a theory for the real understanding of the old methods and also for the search of new methods.

In the first chapter, we unify the various methods for formalizing the notion of sequence transformation. We introduce several definitions that give a system of classes and subclasses of sequence transformations (Cf. the diagram of inclusions on p. ..). We study this system and try to establish the value of each class. In particular, we prove (except in the simplest cases) that the class of algorithmic transformations is strictly included in the class of all transformations. The meaning of this result (and the other results refine this idea), is that some problems are unsolvable with an algorithmic transformation.

In chapter 2, we apply the definitions of chapter 1 to simple problems concerning the transformation of sequences. For example, we study the possibility of "deciding" the convergence of a sequence, of counting the number of accumulation points of a sequence and of determining the period of an asymptotically periodic sequence. Positive results (the definition of algorithms) and negative results (proofs of impossibility) are given.

In chapter 3, we study the problem of extracting convergent sequences from non-convergent sequences. Two kinds of methods are defined and studied : the methods of simultaneous extraction (where several subsequences are built) and the methods of simple extraction. The positive study of the problem of extraction is completed by some negative results.

Chapter 4 is the first of the chapters on the acceleration of convergence. All of the problems of acceleration may be formulated in terms of sequence families. We use this point of view to give a unified presentation of the problems of acceleration. In particular, we study the problem of maximal accelerable families.

Chapter 5 illustrates the utility of the notions of the chapter 1. We give, as precise as possible, a study of the accelerable and non-accelerable families. B. GERMAIN-BONNE and I have introduced a property which is sufficient to prove that a family is non-accelerable. This property (the "remanence") is studied and applied to the families of monotone sequences, the families of linear sequences and the families of logarithmic sequences.

The aim of chapter 6 is to accelerate the convergence of linear sequences. We study the notion of periodico-linear sequences, propose new algorithms for accelerating the convergence of these sequences (the methods for determining the period defined in chapter 2 are used here) and show that in certain senses the Δ^2 of AITKEN is the best transformation for accelerating the family of all linear sequences.

In chapter 7, we present some methods of automatic selection among transformations for accelerating convergence. Theoretical and practical results are given.

REFERENCES

[1]   BREZINSKI C. "Accélération de la convergence en analyse
numérique"
Lecture Notes in Mathematics, 584, Springer—Verlag,
Heidelberg, 1977.

[2]   DENEL J. "Contribution à la synthèse des algorithmes
d'optimisation"
Thèse d'Etat, Lille, 1979.

[3]   GERMAIN—BONNE B. "Estimation de la limite des suites
et formalisation des procédés d'accélératrion de la
convergence"
Thèse d'Etat, Lille, 1978.

[4]   HUARD P. "Optimisation dans R^n :Algorithmes Généraux"
Cours polycopié de D.E.A., Université des Sciences et
Techniques de Lille, 1972.

[5]   PENNACHI R. "Le transformazioni rationali di una
successione"
Calcolo, 5, 1968, pp. 37—50.

[6]   WIMP J. "Sequence transformations and their
applications"
Academic Press, New York, 1981.

[7]   ZANGWILL W. "Nonlinear programming : a unified approach"
Prentice Hall, Englewood Cliffs, 1969.

GENERAL NOTATIONS

N : The set of all integers : $0, 1, \ldots, n \ldots$;

$\mathbf{N}^*$: $\mathbf{N} - \{0\}$;

R : The fields of all real numbers;

$\mathbf{R}^+$: $\{x \in \mathbf{R} \mid x \geq 0\}$;

$\mathbf{R}^*$: $\{x \in \mathbf{R} \mid x \neq 0\}$;

$\mathbf{R}^{+*}$: $\mathbf{R}^+ \cap \mathbf{R}^*$;

C : the field of all complex numbers;

$\mathbf{C}^*$: $\mathbf{C} - \{0\}$

. When A and B are sets

$f : A \to B$: function from A into B ;

dom f   : domain of definition of f
dom $f = \{x \in A \mid f(x) \text{ exists}\}$
if dom f = A , we say that f is an
application

. When E is a set

P(E)   : the set of all the subsets of E

E^n : the set of all the n-uples (finite sequences
with length n) of elements of E .
$E^n = \{(x_1, x_2, \ldots, x_n) \mid \forall \ i \in \{1, \ldots, n\} \ x_i \in E\}$

$E^{(\mathbf{N})}$: the set of all the finite sequences :

$$E^{(\mathbf{N})} = \bigcup_{n \in \mathbf{N}} E^n$$

The empty sequence (of length 0) is denoted
by $\emptyset$

$E^{\mathbf{N}}$: The set of all the infinite sequences of elements
of E .

. When E is a metric space with distance d .

E^α : the set of accumulation points of E :

$E^\alpha = \{x \in E \mid \forall \ \varepsilon \in \mathbf{R}^{+*} , \ \exists \ u \in E : o < d(x,y) \leq \varepsilon\}$

Conv(E)   :   the set of convergent sequence of E^N

Conv*(E)   :   the set of convergent sequences of E^N such
that :

$$\exists\ n_0 \in E \quad \forall\ n \geq n_0 \quad x_n \neq \lim_{k \to \infty} x_k$$

Let $x \in E$, $r \in \mathbf{R}^+$

$$B(x,r) = \{y \in E \mid d(x,y) < r\}$$

Let $A \subset E$, $B \subset E$

$$d(A,B) = \inf\{d(x,y) \mid x \in A\ ,\ y \in B\}$$

$$\delta(A,B) = \sup\{d(x,B)\ ,\ d(A,y) \mid x \in A\ ,\ y \in B\}\ .$$

Chapter 1

The Various Kinds of Algorithmic
Sequence Transformations

INTRODUCTION

In numerical analysis the sequence transformations used are not all of the same kind.

Some of them always use the same formula involving a finite number of terms (for example x_{n-2}, x_{n-1}, x_n) in order to compute the n-th term t_n of the transformed sequence. Some others use a finite constant number of terms but not always the same formula (the formula depends on n) to obtain t_n. Some others use all of the previous $x_0, x_1, \ldots, x_n$, some others change the formula with the sequence (methods of automatic choice are of this type : Cf. [3],[9],[10], chapter 7).

However, from the most specialized to the most general, all sequence transformations satisfy the following principle : the n-th term of the transformed sequence depends only on a finite number of terms of the initial sequence.

The notion of an algorithm for sequences expresses this idea, and we call an algorithmic transformation a sequence transformation for which there exists an algorithm for sequences which computes the transformation.

More simple notions (but also more usable) of algorithm give us particular sequence transformations :

      k-normal transformations, k-memories transformations
      k-stationary transformations, etc...

In this chapter, we are concerned with these definitions and with the general study of these algorithms and transformations. Examples concerning the acceleration of convergence may be found in [1],[2] and [23]. Examples concerning transformations for the extraction of convergent subsequences may be found in [5],[6],[7],[8] and [13].

We concentrate our attention particularly on two problems :

- the problem of classification : what algorithms give the same class of transformations ?
  what inclusions holds for sequence transformations ?
  (propositions 3,4,5,8,11) ;

- the problem of sequence families which may be the domain of definition of a transformation; characterization of such a family; investigation of conditions for a family to be defined everywhere. (propositions 1,2,6,7,9,10,12,13).

Problems of the kernel of regularity and more general problems concerning transformations related to the acceleration of the convergence
are not considered here (see [4],[16],[17],[18],[23] and chapter 4).

This chapter contains many general definitions which are sometimes
complicated. However, we think it essential to have such a set of
notions. On one hand, this set allows us to understand and to classify
the transformations used in acceleration and extraction problems. On
the other hand, these definitions allow us to state and prove negative
results (non-existence of algorithm for certain problems [5],[7],[8],
[11],[12],[13],[14],[15] and [21]).

Most of the definitions given here have been used in other works
(sometimes on slightly different forms) ([1],[5],[7],[11],[12],[14],
[16],[17] and [21]). However, the systematic classification of
sequence transformations and the results on the domains of definition
are new.

Notation

$E^{\mathbf{N}}$: the set of the infinite sequences of elements of E;

$E^{(\mathbf{N})}$: the set of the finite sequences of elements of E;

$$E^{(\mathbf{N})} = \bigcup_{n \in \mathbf{N}} E^n$$

$\emptyset$: the empty set or the empty sequence;

Per(E) : the set of the periodic sequences of elements of E

$$(x_n) \in Per(E) <=> [(x_n) \in E^{\mathbf{N}} \text{ and }$$

$$\exists\, p \in \mathbf{N}^* ,\ \forall\, n \in \mathbf{N} : x_{n+p} = x_n] ;$$

card(E) : the cardinal of E ;

dom(f)   : the domain of definition of the function f ;

Conv(E) :   the set of convergent sequences of E. If $(x_n) \in$ Conv(E)
we denote its limit by x . Similarly, if $(y_n) \in$ Conv(E) we denote
its limit by y , etc...

P(E)   : the set of subsets of E .

1 - SEQUENCE TRANSFORMATIONS

Before we speak about algorithmic sequence transformations, we have to
precisely define a sequence transformation.

Definitions

Let E and F be two sets.

We define a sequence transformations from E^N into F^N to be any function T from the sequence space E^N into the sequence space F^N. We define the domain of definition of T to be the set of sequences (x_n) of E^N for which $T(x_n)$ is defined. This subset of E^N is denoted by dom T .

We say that T is defined on $S \subset E^N$ if dom T $\supset$ S .

We denote the set of all sequence transformations from E^N into F^N by Trans(E,F).

If T and T' are two sequence transformations, we say that T is contained in T' iff :

$$\left| \begin{array}{l} \text{dom T} \subset \text{dom T'} \quad \text{and} \\ \forall \ (x_n) \in \text{dom T} \quad T(x_n) = T'(x_n) \end{array} \right.$$

If $(x_n) \in$ dom T , $T(x_n)$ is a sequence of F^N which is denoted by $(T^{(m)}(x_n))_m$.

If there is no ambiguity about (x_n) , the sequence $T(x_n)$ is denoted by $(T^{(n)})$. If there is no ambiguity about (x_n) or about T , the sequence $T(x_n)$ is denoted by (t_n) .

If there is no ambiguity about T and if the sequences $(x_n{}^0),(x_n{}^1),\ldots,(x_n{}^i),\ldots$ are considered simultaneously (this is frequently the case in proofs of negative results), then the sequences $T(x_n{}^0),T(x_n{}^1),\ldots,T(x_n{}^i),\ldots$ are respectively denoted by

$(t_n{}^0),(t_n{}^1),\ldots,t^i{}_n).$

Examples

1) The identity transformation : T_1 .

   E = F

   $\forall \ (x_n) \in E^N : T_1(x_n) = (x_n)$

   That is to say :

   $\forall \ (x_n) \in E^N \ \forall \ n \in N : t_n = x_n$.

2) Δ^2 transformation of Aitken : T_2 .

   E = F = C

   if $x_{n+2} - 2x_{n+1} + x_n \neq 0$:

$$t_n = (x_{n+2}\, x_n - x^2_{n+1})/(x_{n+2} - 2x_{n+1} + x_n)$$

if not   : $t_n = x_n$

3) Transformation for extraction : T_3 .

E = F metric space; a $\in$ E.

$\forall$ $(x_n) \in E^N$, $t_n = x_i$ where

$i = \max\{j\,|\,n \leq j \leq 2n$, $d(x_j,a) = \min\{d(x_k,a)\,|\,n \leq k \leq 2n\}\}$

4) Transformation for the determination for periodic
    sequences : T_4 .

E ; F = {0,1}

$\forall$ $(x_n) \in E^N$:

$T_4(x_n) = (1,1,\ldots,1,\ldots)$ if (x_n) is periodic

$T_4(x_n) = (0,0,\ldots 0,\ldots)$ if not.

2 - ALGORITHMS FOR SEQUENCES AND ALGORITHMIC TRANSFORMATIONS

A sequence transformation can be effectively used in numerical analysis if it satisfies the following condition (as given in the introduction) :

(*) | The n-th term of the transformed sequence depends
    | only on a finite number of terms of the initial
    | sequence.

The notions of normal transformations ([11],[12],[15]) and of stationary transformations ([16],[17]) express this idea, but there are
transformations in numerical analysis which are not normal, nor stationary (for example T_3). This is why it seems necessary to construct
a definition which includes all of the transformations satisfying (*).

Here is the definition (first given is [5] and then used in
[7],[8],[11],[12],[13],[14] and [15]) :

Definition

An algorithm for sequences from E^N into F^N is :

 (i)   a function **R** : **N** x $E^{(N)}$ x $F^{(N)}$ $\to$ F

(ii)   a function **C** = (α,β) : **N** x $E^{(N)}$ x $F^{(N)}$ $\to$ **N** x **N** .

Applied to the sequence $(x_n) \in E^N$, the algorithm $A = (R,C)$ works in the following way :

Step 0

. Compute $C(0,\emptyset,\emptyset) = (\alpha(0);\beta(0)) \in N \times N$.

. Take the points : $x_{\alpha(0)}, x_{\alpha(0)+1}, \ldots, x_{\beta(0)}$,

. Compute $R(0,(x_{\alpha(0)}, x_{\alpha(0)+1}, \ldots, x_{\beta(0)}),\emptyset) = t_0 \in F$
(t_0 is the first point of the transformed sequence; called the <u>first answer</u>).

. . . .

Step i

. Compute $C(i,(x_{\alpha(i-1)}, x_{\alpha(i-1)+1}, \ldots, x_{\beta(i-1)})$, $(t_0, t_1, \ldots, t_{i-1}))$

$$= (\alpha(i),\beta(i)) \in N \times N$$

. Take the points : $x_{\alpha(i)}, x_{\alpha(i)+1}, \ldots, x_{\beta(i)}$,

. Compute $R(i,(x_{\alpha(i)}, x_{\alpha(i)+1}, \ldots, x_{\beta(i)})$, $(t_0.t_1, \ldots, t_{i-1})) =$

$$= t_i \in F$$

(t_i is the $(i+1)$-th point of the transformed sequence called the <u>$(i+1)$-st answer</u>

. . . .

The set of the sequences $(x_n) \in E^N$ for which this computation can work indefinitely (i.e. always staying in the domain of definition of the functions C and R) is called the domain of definition of the algorithm $A = (R,C)$, and is denoted by dom A .

The set of the algorithms for sequences from E^N into F^N is denoted by $Alg(E,F)$.

If $(x_n) \in$ dom A , we denote by $A(x_n)$ the sequence (t_n) obtained from step $0, \ldots,$ step $n, \ldots$. As with sequence transformations, we may also denote the sequence by $(A^m(x_n))_m$ or (A^n) or (t_n).

From every algorithm $A = (R,C) \in Alg(E,F)$, we obtain a transformation $^T A \in Trans(E,F)$ with domain dom A , which for every sequence $(x_n) \in$ dom(A) gives the sequence $(t_n) \in F^N$ computed by step $0, \ldots,$ step $n, \ldots$

Let $T \in Trans(E,F)$. If there exists $A \in Alg(E,F)$ such that $^T A = T$, we say that T is an algorithmic transformation.

The set of all the algorithmic transformations from E^N into F^N is denoted by : $^\tau Alg(E,F)$.

From the definition : $^\tau Alg(E,F) \subset Trans(E,F)$.

Remark

In the definition, **C** determines the "piece" of the initial sequence which is used in the computation of t_n, and **R** computes this t_n.

It is possible to modify the formulation of our definition without changing the set of algorithmic transformation obtained with the new definition.

For example, we can give a definition where the computation of $(\alpha(n),\beta(n))$ and (t_n) is made using not a section $(x_i, x_{i+1}, \ldots, x_j)$ of the initial sequence but any finite subset of the sequence. We can simplify the definition by saying that t_n is computed with a section $(x_0, x_1, \ldots, x_j)$. We can also choose to not involve previous answers (for they have been obtained from points of the sequence).

Examples

We use the examples of section 1.

1) The identity transformation : T_1 .

   T_1 is algorithmic, indeed $T_1 = {}^\tau A_1$ with $A_1 = (R_1, C_1)$ defined in the following way :

   $\forall\ i \in \mathbf{N}\ ,\ \forall\ s \in E^{(\mathbf{N})}\ ,\ \forall\ s' \in E^{(\mathbf{N})}\ ,\ \forall\ x \in E :$

   $C_1(i,s,s') = (i,i)$,

   $R_1(i,(x),s') = x$,

   C_1, R_1 are not defined elsewhere.

2) Δ^2 transformation of Aitken : T_2

   T_2 is algorithmic; indeed $T_2 = {}^\tau A_2$ where

   $A_2 = (R_2, C_2)$ are defined in the following way :

   $\forall\ i \in \mathbf{N}\ ,\ \forall\ s \in E^{(\mathbf{N})}\ ,\ \forall\ s' \in E^{(\mathbf{N})}\ ;\ \forall\ (x,y,z) \in E^3$

   $C_2(i,s,s') = (i,i+2)$,

   $R_2(i,(x,y,z,),s') = (xz - y^2)/(x - 2y+z)$ if $x - 2y + z \neq 0$,

$$= x \quad \text{if not.}$$

C_2 and R_2 are not defined elsewhere.

3) Transformation for extraction : T_3

T_3 is algorithmic; indeed $T_3 = {}^\tau A_3$ where $A_3 = (R_3, C_3)$ are defined in the following way :

$\forall~i \in \mathbf{N}$, $\forall~s \in E^{(\mathbf{N})}$, $\forall~s' \in E^{(\mathbf{N})}$,

$\forall~(y_0, y_1, \ldots, y_i) \in E^{i+1}$

$C_3(i, s, s') = (i, 2i)$,

$R_3(i, (y_0, y_1, \ldots, y_i), s') = y_\ell$

with $\ell = \max\{j | 0 \leq j \leq i~d(y_j, a) = \min\{d(y_k, a)$

$0 \leq k \leq i\}\}$,

C_3 and R_3 are not defined elsewhere.

4) Transformation for the determination of periodic sequence T_4

A consequence of proposition 4 (of this section) is that T_4 is not algorithmic (if card $E \geq 2$).

This result can be improved in the following way :

If card $E \geq 2$ there is no algorithmic transformation T such that :

$\forall~(x_n) \in \text{Per}(E)~\dashv~n_0 \in \mathbf{N}~\forall~n \geq n_0~~t_n = 1$,

$\forall~(x_n) \notin \text{Per}(E)~\dashv~n_0 \in \mathbf{N}~\forall~n \geq n_0~~t_n = 0$,

$\text{dom}~T = E^{\mathbf{N}}$.

Proposition 1

Let $S \subset E^{\mathbf{N}}$, $S \neq \emptyset$ and let $F \neq \emptyset$:

$(\dashv~T \in {}^\tau \mathbf{Alg}(E, F) : \text{dom}~T = S)$

$\Longleftrightarrow$

$$(\forall \ (x_n) \in E^{\mathbf{N}} : \left[\forall \ k \in \mathbf{N} : \nexists \ (x_n^{\ k}) \in S : \right.$$

$$(x_0, \ldots, x_k) = (x_0^{\ k}, \ldots, x_k^{\ k}) \left.\right] \implies (x_n) \in S)$$

Proof

(i)   Let $T \in {}^{\mathsf{T}}\mathbf{Alg}(E,F)$. We shall show that $S = \text{dom } T$ has the desired property. Let $(x_n) \in E^{\mathbf{N}}$ be such that :

$\forall \ k \in \mathbf{N} , \ \nexists \ (x_n^{\ k}) \in S : (x_0, \ldots, x_k) = (x_0^{\ k}, \ldots, x_k^{\ k})$

Let $A = (\mathbf{R}, \mathbf{C})$ be such that ${}^{\mathsf{T}}A = T$

We shall show by induction that $t_0, \ldots, t_n, \ldots$ are defined.

Let $\mathbf{C}(0, \emptyset, \emptyset) = (\alpha(0), \beta(0))$

Since there exists $(x_n^{\ \beta(0)}) \in S$ such that

$(x_0^{\ \beta(0)}, \ldots, x_{\beta(0)}^{\ \beta(0)}) = (x_0, \ldots, x_{\beta(0)})$,

then $(0, (x_{\alpha(0)}, \ldots, x_{\beta(0)}), \emptyset) \in \text{dom } \mathbf{R}$

and thus t_0 is defined.

Assume that $t_0, t_1, \ldots, t_p$ are defined.

Set $\gamma(p) = \max\{\beta(0), \ldots, \beta(p)\}$.

Since there exists $(x_n^{\ \gamma(p)}) \in S$ such that

$(x_0^{\ \gamma(p)}, \ldots, x_{\gamma(p)}^{\ \gamma(p)}) = (x_0, \ldots, x_{\gamma(p)})$,

then $(p+1, (x_{\alpha(p)}, \ldots, x_{\beta(p)}), (t_0, \ldots, t_p)) \in \text{dom } \mathbf{C}$,

and thus $(\alpha(p+1), \beta(p+1))$ is defined .

Since there exists $(x_n^{\ \gamma(p+1)}) \in S$ such that

$(x_0^{\ \gamma(p+1)}, \ldots, x_{\gamma(p+1)}^{\ \gamma(p+1)}) = (x_0, \ldots, x_{\gamma(p+1)})$,

then $(p+1, (x_{\alpha(p+1)}, \ldots, x_{\beta(p+1)}), (t_0, \ldots, t_p)) \in \text{dom } \mathbf{R}$

and thus t_{p+1} is defined.

Hence (t_n) is defined. That is to say $(x_n) \in S$.

(ii)   Let $S \subset E^{\mathbf{N}}$ satisfy :

$\forall \ (x_n) \in E^{\mathbf{N}} :$

$$\left[\forall\ k \in \mathbf{N}\ ,\ \exists\ (x_n{}^k) \in S\ (x_0, \dots, x_k) = (x_0{}^k, \dots, x_k{}^k) \right]$$

$$\Rightarrow (x_n) \in S\ .$$

Let $f \in F$. We define $A = (R,C)$ by :

$\forall\ k \in \mathbf{N},\ \forall\ (x_n) \in S$:

$$C(k, (x_0, \dots, x_{k-1})\ ,\ \underbrace{(f, \dots, f)}_{k \text{ times}}) = (0, k)\ \ ,$$

$$R(k, (x_0, \dots, x_k)\ ,\ \underbrace{(f, \dots, f)}_{k \text{ times}}) = f\ ,$$

R and C are not defined elsewhere.

We have : dom $C = \{(k, s, s') \in \mathbf{N} \times E^{\mathbf{N}} \times F^{(\mathbf{N})} | \exists\ (x_n) \in S$

$s = (x_0, \dots, x_{k-1})$ $s' = \underbrace{(f, \dots, f)}_{k \text{ times}}\}$

dom $R = \{(k, s, s') \in \mathbf{N} \times E^{(\mathbf{N})} \times F^{(\mathbf{N})} | \ \exists\ (x_n) \in S$

$s = (x_0, \dots, x_k)$ $s' = \underbrace{(f, \dots, f)}_{k \text{ times}}\}$

Obviously $S \subset$ dom A .

Let $(x_n) \notin S$. From the hypothesis :

$\exists\ k \in \mathbf{N}\ ,\ \forall\ (y_n) \in S : (x_0, \dots, x_k) \neq (y_0, \dots, y_k)$.

Thus $(k, (x_0, \dots, x_k)\ ,\ \underbrace{(f, \dots, f)}_{k \text{ times}}) \notin$ dom R

Hence dom $A \subset S$.

Remarks

1) The characterisation given in proposition 1 may be formula-
    ted as follows :

    S is a closed set of $E^{\mathbf{N}}$ (when the topology on $E^{\mathbf{N}}$ is
    the product topology of the discrete topology on E) .

2) If $E = \mathbf{R}$, then the set of convergent sequences
    (resp. linearly convergent sequences, resp. logarithmi-
    cally convergent sequences) does not satisfy the charac-
    tisation of the families which are the domain of defini-
    tion of an algorithmic transformation. This implies that
    the transformations (for example those for accelerating
    convergence) which give answers for every convergent
    sequence also give answers for certain non-convergent
    sequences.

Let $S \subset E^{\mathbf{N}}$. We denote by $\overline{S}$ the subset of $E^{(\mathbf{N})}$ containing all the
beginnings of sequences of S :

$$\overline{S} = \{(x_0,\ldots,x_p) \mid (x_n) \in S\} \subset E^{(\mathbf{N})}$$

Here is a consequence of proposition 1 :

Proposition 2

Let $T \in {}^{\tau}\mathrm{Alg}(E,F)$. Let $S = \mathrm{dom}\ T$:

$$\overline{S} = E^{(\mathbf{N})} \iff S = E^{\mathbf{N}}$$

This proposition is important, for it allows us to prove that (except
in obvious cases) there always exist transformations which are not
algorithmic (i.e. our definition on page 13 is genuinely restrictive).

Proposition 3

$({}^{\tau}\mathrm{Alg}(E,F) = \mathrm{Trans}(E,F))$

$\iff$

(card $E < 2$ or card $F < 1$)

Proof

a) We assume that card = 0 or card F = 0 .
    Then Trans(E,F) contains only the function with an
    empty domain, which is algorithmic (define $A = (R,C)$
    with dom$(R) = $ dom$(C) = \emptyset$).

b) We assume that card E = 1 , $E = \{e\}$.
    Let $T \in$ Trans(E,F) . Either dom$(T) = \emptyset$ (and
    then T is algorithmic) or
    dom$(T) = \{(e,e,\ldots,e,\ldots)\}$
    $E^{\mathbf{N}} = \{(e,e,\ldots,e,\ldots)\}$ and then let
    $(t_n) = T(e,e,\ldots,e,\ldots)$
    we define $A = (R,C)$:

$\forall\ i \in \mathbf{N}$

$$C(i,\underbrace{(e,e,\ldots,e)}_{i\ times}\ ,\ (t_0,t_1,\ldots,t_{i-1})) = (0,i)$$

$$R(i,\underbrace{(e,e,\ldots,e)}_{(i+1)\ times}\ ,\ (t_0,t_1,\ldots,t_{i-1})) = t_i$$

$\mathbf{R},\mathbf{C}$ not defined elsewhere.

Obviously, $^\tau A = T$, so that T is algorithmic.

c) We assume that : $card(F) \geq 1$ and $card\ E \geq 2$
We have :

$$Per(E) \neq E^{\mathbf{N}}$$

$$\overline{Per}(E) = E^{(\mathbf{N})}$$

and thus if T is a transformation having the domain of definition $Per(E)$, so that T is not algorithmic (proposition 2).

Proposition 4

$(\forall\ T \in Trans(E,F)\ \dashv\ T' \in\ ^\tau Alg(E,F) : T \subset T')$

$\qquad \langle = \rangle$

$(card\ E < 2\quad or\quad card\ F < 2)$

Proof

a) If $card\ E < 2$ or if $card\ F < 1$, the result follows from Proposition 3.

b) If $card\ E = 1$, then every transformation T is contained in the transformation T' with domain $E^{\mathbf{N}}$ and definde by :

$\forall(x_n) \in E^{\mathbf{N}}\ :\ T(x_n) = (f,f,\ldots,f,\ldots)$

$(F = \{f\})$

The transformation T' is algorithmic : $T = {}^\tau A$, with

$A = (\mathbf{R},\mathbf{C})$ defined by :

$\forall\ i \in \mathbf{N}\ ,\ \forall\ s \in E^{(\mathbf{N})}\ ,\ \forall\ s' \in F^{(\mathbf{N})}$:

$C(i,s,s') = (0,i)$

$R(i,s,s') = f$

c) We assume that : card E $\geq$ 2 and card F $\geq$ 2

Let $f_1 \in F$, $f_2 \in F$, $f_1 \neq f_2$.

Let   T be the following transformation with domain E^N :

$T(x_n) = (f_1,f_1,\ldots,f_1,\ldots)$ if $(x_n) \in Per(E)$

$T(x_n) = (f_2,f_2,\ldots,f_2,\ldots)$ if $(x_n) \notin Per(E)$

If T is contained in an algorithmic transformation, then
T is also algorithmic (for dom T = E^N).
Thus there exists A = (R,C) such that $^\tau A = T$.
Let $(\alpha(0),\beta(0)) = C(0,\emptyset,\emptyset) \in N^2$.
Since card E $\geq$ 2 , there exists (x^0_n) , (x^1_n)
such that :

(i) $(x^0_n) \in Per(E)$,

(ii) $(x^1_n) \notin Per(E)$,

(iii) $x^0_n = x^1_n$ for every $n \in \{\alpha(0),\alpha(0)+1,\ldots,\beta(0)\}$.

From (i) and (ii) , we obtain $T^{(0)}(x^0_n) = f_1$,

$$T^{(0)}(x^1_n) = f_2 .$$

From (iii) , we obtain $A^0(x^0_n) = A^0(x^1_n)$.

Hence it is impossible that $^\tau A = T$.

Remark

Propositions 3 and 4 mean the following :

If E and F are not too small, then there exist transformations
which are not algorithmic, and even which cannot match any algorithmic
transformation.

3 - k-NORMAL ALGORITHMS, k-NORMAL TRANSFORMATIONS

In the acceleration of convergence, it is important to consider only
the sequence transformations which use the points :

$$x_0,x_1,\ldots,x_n,x_{n+1},\ldots,x_{n+k}$$

for the computation of t_n , where k is a fixed integer (with a shift of the indices, it is possible ot consider only the case where t_n is computed using $x_0, x_1, \ldots, x_n$) .

This is why we define the notion of a k-normal algorithm. As algorithms give rise to algorithmic transformations, here we obtain k-normal transformations.

Every k-normal transformation is an algorithmic transformation. The converse is not true (proposition 3). Less obvious : the families of sequences which are the domain of definition of a k-normal transformation are the same families which are the domain of definition of an algorithmic transformation (proposition 6).

Definitions and notations

Let $k \in \mathbf{N}$.

We call a k-normal algorithm from $E^{\mathbf{N}}$ into $F^{\mathbf{N}}$ any sequence of functions (f_n) such that :

$$\forall\, n \in \mathbf{N} \qquad f_n : E^{n+k+1} \to F$$

Applied to the sequence $(x_n) \in E^{\mathbf{N}}$, the k-normal algorithm $A = (f_n)$ works in the following way :

$$t_0 = f_0(x_0, \ldots, x_k)$$

$$t_1 = f_1(x_0, \ldots, x_{1+k})$$

$$\ldots$$

$$t_n = f_n(x_0, \ldots, x_{n+k})$$

$$\ldots$$

When $k = 0$, rather than "k-normal algorithm" we refer to <u>normal algorithm</u>.

The domain of definition of the algorithm $A = (f_n)$ (denoted by dom A) is the set of the sequences $(x_n) \in E^{\mathbf{N}}$ such that :

$$\forall\, n \in \mathbf{N} \; : \; (x_0, x_1, \ldots, x_{n+k}) \in \text{dom } f_n \; .$$

The set of all k-normal algorithms from $E^{\mathbf{N}}$ into $F^{\mathbf{N}}$ is denoted by $\text{Norm}_k(E,F)$.

From every algorithm $A \in \text{Norm}_k(E,F)$, we obtain a transformation $^\tau A \in \text{Trans}(E,F)$ having dom A as domain of definition and defined by

$$\forall\, n \in \mathbf{N} \quad t_n = f_n(x_0, \ldots, x_n, \ldots, x_{n+k})$$

Let $T \in \mathbf{Trans}(E,F)$. If there exists $A \in \mathbf{Norm}_k(E,F)$ such that $T = {}^\tau A$, we say that T is a k-normal transformation.

The set of all k-normal transformations from E^N into F^N is denoted by ${}^\tau \mathbf{Norm}_k(E,F)$.

From the definition, we have ${}^\tau \mathbf{Norm}_k(E,F) \subset \mathbf{Trans}(E,F)$ but if, as we claim, our notion of algorithmic transformation is sufficiently general, we must have ${}^\tau \mathbf{Norm}_k(E,F) \subset {}^\tau \mathbf{Alg}(E,F)$. This is established in proposition 5 .

Examples (see § 1)

1) The identity transformation : T_1.

T_1 is a normal transformation for $T_1 = {}^\tau N_1$, where $N_1 = (f_n)$:

$$\forall\ n \in \mathbf{N}\ ,\ \forall\ x_0,\dots,x_n \in E\ :\ f_n(x_0,\dots,x_n) = x_n$$

2) Δ^2 transformation of Aitken : T_2.

T_2 is a 2-normal transformation for $T_2 = {}^\tau N_2$, where $N_2 = (f_n)$

$$f_n(x_0,x_1,\dots,x_{n+2}) = (x_{n+2}\ x_n - x^2_{n+1})/(x_{n+2} - 2\ x_{n+1} + x_n)$$

if $x_{n+2} - 2\ x_{n+1} + x_n \neq 0$,

$$= 0\ \text{otherwise.}$$

It is possible to "normalize" the transformation T_2 by setting :

$t'_n = x_n$ if $n = 0$ or $n = 1$

$t'_n = (x_n\ x_{n-2} - x^2_{n-1})/(x_n - 2\ x_{n-1} + x_{n-2})$ if

$\quad x_n - 2\ x_{n-1} + x_{n-2} \neq 0$ and $n \geq 2$,

$t'_n = x_n$ if $x_n - 2\ x_{n-1} + x_{n-2} = 0$ and $n \geq 2$.

This new transformation T'_2 is called the normalized Δ^2 transformation. Obviously T'_2 is normal (i.e. 0-normal).

When we define a transformation by a formula such as $t_n = h_n(x_{n-p},\dots,x_p)$, it is implicitly assumed that the associated normal algorithm is $A = (f_n)$:

$t_n = x_n = f_n(x_0,\dots,x_n)$ if $n < p$

$t_n = h_n(x_{n-p},x_{n-p+1},\dots,x_n) = f_n(x_0,\dots,x_n)$ if $n \geq p$.

T_3 and T_4 are not normal transformations (nor k-normal transformation, for any k).

For T_3 , this is shown by contradiction (in every metric space containing b,c with $d(a,b) \neq d(a,c)$).
Concerning T_4 , the result is a consequence of proposition 5 (because T_4 is not algorithmic).

Proposition 5

i) $\forall k \in \mathbf{N}$: $^\tau Norm_k(E,F) \subset {}^\tau Norm_{k+1}(E,F) \subset {}^\tau Alg(E,F)$

$$\bigcup_{k \in \mathbf{N}} {}^\tau Norm_k(E,F) \subset {}^\tau Alg(E,f)$$

ii) there are equalities if and only if :

$$card(E) < 2 \quad or \quad card(F) < 2$$

Proof

a) The first inclusion is obvious.

b) We show the second. Let $A = (f_n) \in Norm_k(E,F)$.
We define $B = (R,C) \in Alg(E,F)$:

$$C(0,\emptyset,\emptyset) = (0,k) \quad and$$

$$\forall (x_n) \in dom \ A \ , \ \forall n \in N :$$

$C(n,(x_0,\dots,x_{n+k-1}),(f_0(x_0,\dots,x_k),\dots,$
$$f_{n-1}(x_0,\dots,x_{n+k-1}))) = (0,n+k)$$

$R(n,(x_0,\dots,x_{n+k}),(f_0(x_0,\dots,x_k),\dots,$
$$f_{n-1}(x_0,\dots,x_{n+k-1}))) = f_n(x_0,\dots,x_{n+k})$$

R and C are not defined elsewhere.

We verify that dom B = dom A and A = B.

c) The third inclusion is easy to obtain from the second.

d) When card E < 2 or card F < 2 , it is not difficult to see that inclusions become equalities.

e) Assume that card $E \geq 2$ and card $F \geq 2$.
The transformation "k+1 shift"

$$T : (x_n) \to (t_n) \ ; \ t_n = x_{n+k+1}$$

is a tranformation of $^\tau Norm_{k+1}(E,F)$ and a simple reasoning by contradiction shows that $T \notin {}^\tau Norm_k(E,F)$.

Similarly, the transformation :

$$T : (x_n) \to (t_n) \; ; \; t_n = x_{2n} \quad,$$

does not belong to $^\tau Norm_k(E,F)$ (for every k) , but is an algorithmic transformation.

Remark

The set of the transformations contained in an algorithmic transformation and the set of the transformations contained in a k-normal transformation are equal if and only if :

$$card(E) < 2 \quad or \quad card(F) < 2.$$

("<=" is obvious. To prove "=>" we use the two transformations of proof of part (e) of proposition 5) .

The families of sequences which are the domain of definition of a k-normal transformation (from proposition 5) are also the domain of definition of an algorithmic transformation. It is remarkable that the converse is also true :

Proposition 6

Let $S \subset E^N$: $(\dashv T \in {}^\tau Norm_k(E,F) : dom\ T = S)$

$$<=>$$

$$(\dashv T \in {}_\tau Alg(E,F) : dom\ T = S)$$

Proof

. => Proposition 5

. <= Let $S \subset E^N$ be such that there exists $T \in {}^\tau Alg(E,F)$ satisfying $dom\ T = S$.

The case $F = \emptyset$ is obvious. Hence, let $f \in F$. We define

$N = (f_n) \in Norm_k(E,F)$ by :

$$\forall\ n \in N\ \forall\ (x_n) \in S : f_n(x_0,\ldots,x_{n+k}) = f \ .$$

Proposition 1 shows that $dom\ N = dom\ T$.

From propositions 2 and 6, we obtain :

Proposition 7

Let $T \in {}^{\tau}\mathrm{Norm}_k(E,F)$. Let $S = \mathrm{dom}\ T$;

$\overline{S} = E^{(\mathbf{N})} \iff S = E^{\mathbf{N}}$.

4 — k-MEMORIES ALGORITHMS AND k-MEMORIES TRANSFORMATIONS

Sometimes the computation of t_n depends only on n and on the terms $x_{n-k+1}, \ldots, x_n$. This kind of algorithm is very important in the acceleration of convergence, which is why we study it and give it a special name.

Of course, a k-memories transformation is a normal transformation (proposition 8), but now there are less families which are the domain of definition of a k-memories transformation than families which are the domain of definition of an algorithmic transformation (proposition 9).

Definitions and notation.

Let $k \in \mathbf{N}$.

We call a k-memories algorithm from $E^{\mathbf{N}}$ into $F^{\mathbf{N}}$ any sequence of functions (f_n) such that :

$$\forall\ n < k \qquad f_n : E^{n+1} \to F$$

$$\forall\ n \geq k-1 \qquad f_n : E^k \to F$$

Applied to the sequence $(x_n) \in E^{\mathbf{N}}$, the k-memories algorithm $M = (f_n)$ works in the following way :

$$(1) \quad \left|\begin{array}{ll} t_n = f_n(x_0, x_1, \ldots, x_n) & \text{if } n < k , \\[2mm] t_n = f_n(x_{n-k+1}, \ldots, x_n) & \text{if } n \geq k-1 . \end{array}\right.$$

When $E = F$, without special definition, we always set

$$\forall\ n < k-1 \quad f_n(x_0, \ldots, x_n) = x_n .$$

The set of sequences $(x_n) \in E^{\mathbf{N}}$ such that

$$\forall\ n < k \quad (x_0, x_1, \ldots, x_n) \in \mathrm{dom}\ f_n ,$$

$$\forall\ n \geq k-1 \quad (x_{n-k+1}, \ldots, x_n) \in \mathrm{dom}\ f_n ,$$

is the domain of definition of $M = (f_n)$, this set being denoted by dom T .

The set of all the k-memories algorithms is denoted by $\mathbf{Mem}_k(E,F)$.

From every k-memories algorithm $M \in \mathbf{Mem}_k(E,F)$, we obtain a transformation $^\tau M \in \mathrm{Trans}(E,F)$ having dom M as domain of definition and defined by (1) .

Let $T \in \mathrm{Trans}(E,F)$. If there exists $M \in \mathbf{Mem}_k(E,F)$ such that $^\tau M = T$, we say that T is a k-memories transformation.

The set of all the k-memories transformations from E^N into F^N is denoted by $^\tau\mathbf{Mem}_k(E,F)$.

Obviously, $^\tau\mathbf{Mem}_k(E,F) \subset \mathrm{Trans}(E,F)$.

Example (see § 1,2,3)

T_1 is a 1-memory transformation.

T'_2 is a 3-memories transformation.

But T_2, T_3, T_4 are not k-memories transformations for any k .

Proposition 8

i) $\forall k \in \mathbf{N}$, $\forall p \in \mathbf{N}$

$^\tau\mathbf{Mem}_p(E,F) \subset {}^\tau\mathbf{Mem}_{p+1}(E,F) \subset {}^\tau\mathbf{Norm}_k(E,F)$

$\underset{j \in \mathbf{N}}{\cup} \ {}^\tau\mathbf{Mem}_j(E,F) \subset {}^\tau\mathbf{Norm}_k(E,F)$

ii) There are equalities if and only if

card E < 2 or card F < 1 .

Proof

i) Obvious.

ii) If card E < 2 or card F < 1 , then it is easy to see that they are equalities.

We assume that card $E \geq 2$ and card $F \geq 1$. Let $f \in F$. We define a transformation T :

$T(x_n) = (f,f,\ldots,f,\ldots)$ if (x_n) satisfies :

$$\forall\ n \in \mathbf{N}\quad x_n = x_{2n}\ ,$$

$T(x_n)$ is not defined elsewhere.

T is a normal transformation for $T = {}^{\tau}A$ with

$$A = (f_n) \in \mathbf{Norm}(E)$$

is defined by :

$$\forall\ n \in \mathbf{N}\quad \forall\ (x_0, x_1, \ldots, x_n) \in E^{n+1}$$

$(\forall\ p : 2p \leq n \Rightarrow x_p = x_{2p}) \Rightarrow \left[(x_0, \ldots, x_n) \in \text{dom}\ f_n \right.$
and $\left. f_n(x_0, \ldots, x_n) = f\right]$

In order to show that T is not a k-memories transformation, we assume that $T = {}^{\tau}M$ with $M = (g_n) \in \mathbf{Mem}_k(E, F)$.

Let k' be odd, $k' \geq k$. Let $e_1, e_2 \in E$, $e_1 \neq e_2$.

We define (x_n) and (y_n) :

$x_n = e_1$ if there is $i \in \mathbf{N}$ such that $n = 2^i k'$,

$x_n = e_2$ otherwise,

$y_n = e_2$ for every $n \in \mathbf{N}$.

We have $(x_n) \in \text{dom}\ M$, $(y_n) \in \text{dom}\ M$. Consequently, the sequence (z_n) defined by :

$z_{k'} = e_1$, $z_n = e_2$ for every $n \neq k'$,

is a sequence of dom M .

But $(z_n) \notin \text{dom}\ T$, hence ${}^{\tau}M \neq T$.

We have just proved that, if card $E \geq 2$ and card $F \geq 1$, then there exist normal transformations which are not k-memories transformations.

A similar argument shows that the following transformation:

$T(x_n) = (f, f, \ldots, f, \ldots)$ if $\forall\ n \in \mathbf{N}\quad x_n = x_{n+k}$,

$T(x_n)$ not defined otherwise

is a (k+1)-memories transformation, but not a k-memories transformation.

Proposition 9

Let $k \in \mathbf{N}$, $\mathbf{S} \subset E^{\mathbf{N}}$.

If $F \neq \emptyset$, then :

$$(\exists\, T \in {}^{\tau}\mathrm{Mem}_k(E,F) : \mathrm{dom}\, T = \mathbf{S})$$

$$\Longleftrightarrow$$

$$\left[\forall\, (x_n) \in E^{\mathbf{N}}\, (\forall\, i \in \mathbf{N}\,, \exists\, (x^i_n) \in \mathbf{S} :\right.$$

$$\left.(x^i_i, x^i_{i+1}, \ldots, x^i_{i+k-1}) = (x_i, x_{i+1}, \ldots, x_{i+k-1})) \Rightarrow (x_n) \in \mathbf{S}\right].$$

Proof

. $\Rightarrow$.

Let $M = (f_n) \in \mathrm{Mem}_k(E,F)$ be such that $\mathrm{dom}\, M = \mathbf{S}$.
Let $(x_n) \in E^{\mathbf{N}}$ be such that :

$\forall\, i \in \mathbf{N}\,, \exists\, (x^i_n) \in \mathbf{S} :$

$$(x^i_i, x^i_{i+1}, \ldots, x^i_{i+k-1}) = (x_i, x_{i+1}, \ldots, x_{i+k-1})$$

With $i = 0$, we obtain :

$\forall\, n \leq k-1 : (x_0, x_1, \ldots, x_n) \in \mathrm{dom}\, f_n$.

With $i \in \mathbf{N}$, we obtain :

$\forall\, n \geq k-1 : (x_{n-k+1}, \ldots, x_n) \in \mathrm{dom}\, f_n.$

Thus $(x_n) \in \mathbf{S}.$

. $\Leftarrow$.

Let $\mathbf{S}$ satisfy $[\ldots]$. Let $f \in F$.

We define $M = (f_n)$:

$\forall\, (x_n) \in \mathbf{S}\ \ \forall\, m \leq k-1\ \ f_m(x_0, \ldots, x_m) = f$,

$\qquad\qquad \forall\, m \geq k-1\ \ f_m(x_{m-k+1}, \ldots, x_m) = f,$

f_n is not defined elsewhere.

From the construction, $\mathrm{dom}\, M \supset \mathbf{S}$, the hypothesis on $\mathbf{S}$ shows that $\mathrm{dom}\, M \subset \mathbf{S}$

Proposition 10

Let $T \in {}^{\tau}Mem_k(E,F)$.

Let $S = \text{dom } T$.

If $F \neq \emptyset$, then :

$$\left[\forall \, i \in \mathbf{N} : \, \{(x_i,\ldots,x_{i+k-1}) \,|\, (x_n) \in S\} = E^k\right]$$

$$<=>$$

$$\left[\overline{S} = E^{(\mathbf{N})}\right]$$

$$<=>$$

$$\left[S = E^{\mathbf{N}}\right]$$

5 - k-STATIONARY ALGORITHMS AND k-STATIONARY TRANSFORMATIONS

In the acceleration of convergence, many algorithms use only $x_{n-k-1},\ldots,x_n$ in order to compute t_n , without involving the step number n . We now study those transformations, which according to the usual terminology ([16]) are called stationary transformations.

Definitions and notations

Let $k \in \mathbf{N}^*$. We call a k-stationary algorithm from $E^{\mathbf{N}}$ into $F^{\mathbf{N}}$ any k-memories algorithm (f_n) such that :

$$\forall \, n \geq k-1 \quad f_n = f_{k-1} \; .$$

When $E = F$, without special definition, we always set :

$$\forall \, n \leq k-1 \; , \; \forall \; (x_0,\ldots,x_n) : f_n(x_0,\ldots,x_n) = x_n$$

In such a situation, a k-stationary algorithm is determined by one function $f : E^k \to E$ and the transformed sequence of (x_n) is defined by :

(2) $t_n = x_n$ if $n < k-1$; $t_n = f(x_{n-k+1},\ldots,x_n)$ if $n \geq k-1$.

The set of all k-stationary algorithms is denoted by $Stat_k(E,F)$.

From every k-stationary algorithm $A \in Stat_k(E,F)$, we obtain a transformation ${}^{\tau}A \in Trans(E,F)$ having dom A as domain of definition and defined by (2).

Let $T \in Trans(E,F)$. If there exists $A \in Stat_k(E,F)$ such that ${}^{\tau}A = T$ we say that T is a k-stationary transformation.

The set of all the k-stationary transformations is denoted by : ${}^{\tau}Stat_k(E,F)$.

From the definitions :

$$^\tau Stat(E,F) \subset Trans(E,F).$$

Examples (see § 1,2,3,4)

T_1 is a 1-stationary transformation.

T'_2 is a 3-stationary transformation.

T_2,T_3 and T_4 are not k-stationary (for any k)

Proposition 11

$\forall\ k \in \mathbf{N}^*$

(i) $^\tau Stat_k(E,F) \subset {}^\tau Stat_{k+1}(E,F) \subset {}^\tau Mem_{k+1}(E,F).$

(ii) There are equalities if and only if :

$$card\ E < 2\quad or\quad card\ F < 1 \ .$$

Proof

(i)   Obvious

(ii) If card E < 2 or card F < 1 , then of course there are equalities.

Assume that   card E $\geq$ 2   and   card F $\geq$ 1 .

The transformation defined by :

$T(x_n) = (f,f,\ldots,f,\ldots)$ if, for every   n   even :

$x_n = x_{n+k-1}$ and for every   n odd : $x_n \neq x_{n+k-1}$,

$T(x_n)$ no defined elsewhere,

is a k-memories transformation but not a k-stationary transformation.

The transformation defined by

$T(x_n) = (f,f,\ldots,f,\ldots)$ if for every n $x_n = x_{n+k}$,

$T(x_n)$ no defined elsewhere,
is a   (k+1)-stationary transformation but not a k-stationary transformation.

Proposition 12

Let $S \subset E^{\mathbf{N}}$.

If there exists $S \in {}^{\tau}Stat_k(E,F)$ such that dom $S = S$, then :

$\exists\, A \subset E^k$, $\forall$ $(x_n) \in S$, $\forall$ $i \in N$ $(x_i,\ldots,x_{i+k-1}) \in A$.

Proposition 13

Let $S \in {}^{\tau}Stat_k(E,F)$ and let $S = $ dom S .

$$\left[\{(x_0,\ldots,x_{k-1})|(x_n) \in S\} = E^k\right] <=>$$

$$\left[\overline{S} = E^{(\mathbf{N})}\right] <=>$$

$$\left[S = E^{\mathbf{N}}\right] .$$

6 — RATIONAL TRANSFORMATIONS, LINEAR TRANSFORMATION

Assume that $E = F = \mathbf{K}$, where $\mathbf{K}$ is a field.

We say that the k-normal algorithm $N = (f_n)$ is _rational_ if, for every $n \in N$, f_n is a rational function with respect to the variables $x_0, x_1, \ldots, x_{n+k}$.

We denote by $Norm^r_k(E,F)$ the set of all the k-normal rational algorithms.

Let $T \in Trans(E,F)$. If there exists a k-normal rational algorithm $N = (f_n)$ such that ${}^{\tau}N = T$, we say that T is a k-normal rational transformation.

We denote by ${}^{\tau}Norm^r_k(E,F)$ the set of all the k-normal rational transformations.

Similarly, we define :

The k-memories rational algorithms ;
The k-stationary rational algorithms;
The k-memories rational transformations;
The k-stationary rational transformations.

We introduce the notation :

$Mem^r_k(E,F)$, ${}^{\tau}Mem^r_k(E,F)$, $Stat^r_k(E,F)$, ${}^{\tau}Stat^r_k(E,F)$.

Assume that E and F are linear spaces on $\mathbf{K}$.

We say that the k-normal algorithm $\mathbf{N} = (f_n)$ is _linear_ if, for eveny $n \in N$, f_n is a linear transformation from E^{n+k+1} into F .

As before, we define :

The k-memories linear algorithms;
The k-stationary linear algorithms;
The k-memories linear transformations;
The k-stationary linear transformations.

We introduce the notation :

$\text{Norm}^{\ell}_{k}(E,F)$, $^{\tau}\text{Nomr}^{\ell}_{k}(E,F)$, $\text{Mem}^{\ell}_{k}(E,F)$, $^{\tau}\text{Mem}^{\ell}_{k}(E,F)$,

$\text{Stat}^{\ell}_{k}(E,F)$, $^{\tau}\text{Stat}^{\ell}_{k}(E,F)$.

Examples

The transformations T_1, T_2, T'_2 are rational transformations. Almost all of the transformations used in the acceleration of convergence are rational (when normalized). This is the case for the ε , ρ and θ-algorithms. However, the algorithms obtained by selection (chapter 7) are not rational.

7 - INCLUSION DIAGRAM

When $E = F = \mathbf{K}$, we obtain :

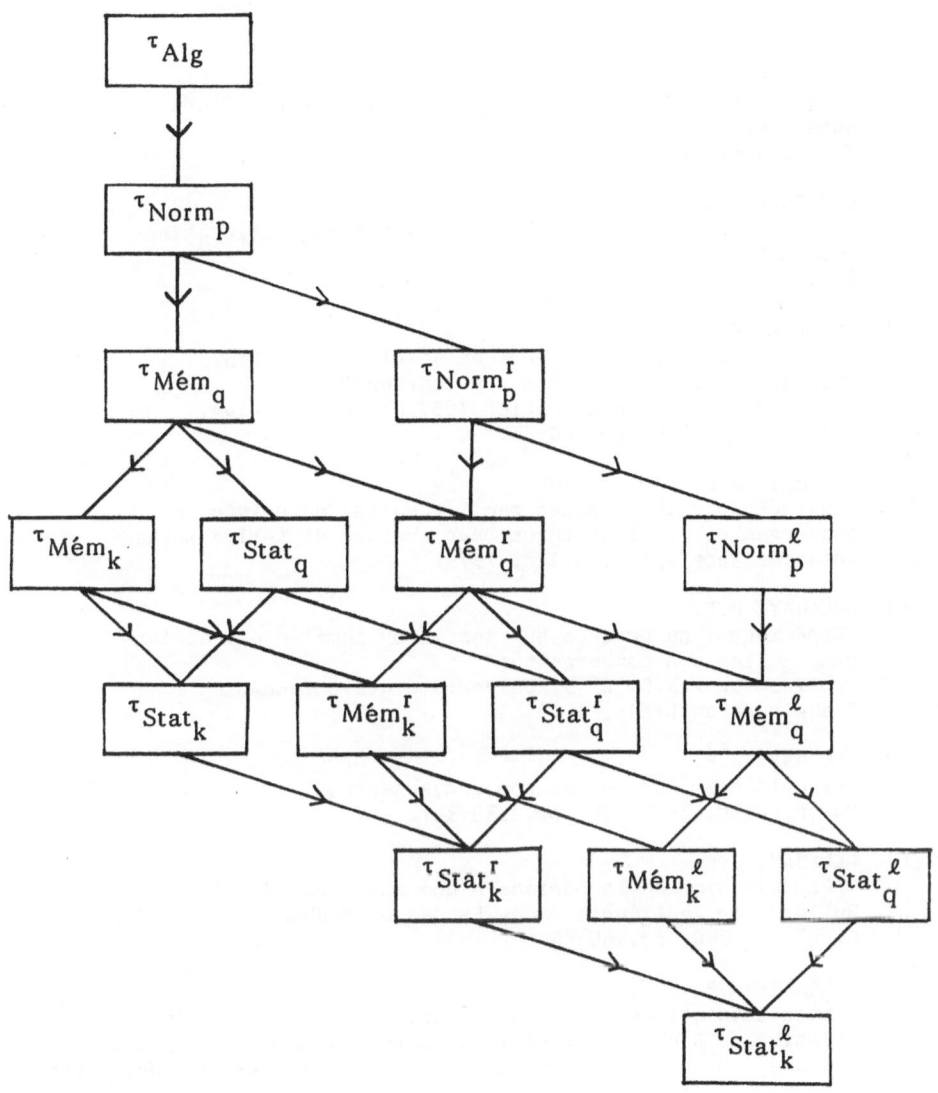

REFERENCES

[1] BREZINSKI C.
     "Accélération de la convergence en Analyse Numérique"
     Lectures Notes in Mathematics 584, Springer-Verlag,
     Heidelberg, 1977.

[2] BREZINSKI C.
     "Algorithmes d'accélération de la convergence : Etude
     Numérique".
     Technip, Paris, 1978

[3] BREZINSKI C.
     "Asymptotic error estimates in convergence acceleration
     methods"
     To appear.

[4] CORDELLIER F.
     "Sur la régularité des procédés δ^2 et w de Lubkin.
     Padé approximation and its applications"
     Lectures Notes in Mathematics 765, Springer-Verlag,
     Heidelberg, 1980, pp. 20-35.

[5] DELAHAYE J.P.
     "Quelques problèmes posés par les suites de points non
     convergentes et algorithmes pour traiter de telles suites"
     Thèse de 3ème Cycle, Lille, 1979.

[6] DELAHAYE J.P.
     "Expériences numériques sur les algorithmes d'extraction
     pour suites non convergentes"
     Publication A.N.O. n° 5, Université des Sciences et
     Techniques de Lille, 1979.

[7] DELAHAYE J.P.
     "Algorithmes pour suites non convergentes"
     Numer. Math. 34, 1980, pp. 333-347.

[8] DELAHAYE J.P.
     "Détermination de la période d'une suite pseudo-périodique"
     Bulletin de la Direction des Etudes et Recherches de l'E.D.F.,
     C, n° 1, 1980, pp. 65-80.

[9] DELAHAYE J.P.
     "Choix automatique entre suites de paramètres dans l'extra-
     polation de Richardson. Padé approximation and its applications",
     Lecture Notes in Mathematics 888, Springer-Verlag, Heidelberg,
     1981, pp. 158-172.

[10] DELAHAYE J.P.
     "Automatic selection of sequence transformations"
     Math. Comp., 37, 1981, pp. 197-204.

[11] DELAHAYE J.P.
"Optimalité du procédé Δ^2 D'Aitken pour l'accélération
de la convergence linéaire"
RAIRO Analyse Numérique, 15, 1981; pp. 321–330.

[12] DELAHAYE J.P.
"Accélération des suites dont le rapport des erreurs est
borné",
Calcolo, 18, 1981, pp. 103–116.

[13] DELAHAYE J.P.
"Algorithmes pour extraire une sous-suite convergente
d'une suite non convergente"
Conference Optimization : Theory and Algorithm,
16–20 Mars 1981, Lecture Notes in Pure and Applied
Mathematics, Marcel Dekker, New York.

[14] DELAHAYE J.P. et GERMAIN-BONNE B.
"Résultats négatifs en accélération de la convergence"
Numer. Math., 35, 1980, pp. 443–457.

[15] DELAHAYE J.P. et GERMAIN-BONNE B.
"The set of logarithmically convergent sequences cannot
be accelerated"
SIAM, J. Numer. Anal., 19, 1982, pp. 840–844.

[16] GERMAIN-BONNE B.
"Estimation de la limite de suites et formalisation des
procédés d'accélération de convergence",
Thèse, Lille, 1978.

[17] GERMAIN-BONNE B.
"Transformations de suites"
RAIRO, R-1, 1973, pp. 84–90.

[18] GERMAIN-BONNE B.
"Conditions suffisantes d'accélérabilité"
Publication ANO, n° 32, Université des Sciences et Techniques
de Lille, 1981.

[19] KOWALEWSKI C.
"Possibilités d'accélération de la convergence logarithmique"
Thèse de 3ème cycle, Lille, 1981.

[20] KOWALEWSKI C.
"Accélération de la convergence pour certaines suites à
convergence logarithmique"
Padé approximation and its Applications
Lectures Notes in Mathematics 888, Springer-Verlag, Heidelberg,
1981, pp. 263–272.

[21] PENNACHI R.
"Le transformazioni rationali di una successione"
Calcolo, 5, 1968, pp. 35-50.

[22] TROJAN J.M.
"An upper bound on the acceleration of convergence"
First French-Polish Meeting on Padé Approximation and Convergence
Acceleration Techniques. Varsovie, Pologne, 1-4 Juin 1981.

Chapter 2
Decidability and Undecidability
in the Limit

INTRODUCTION

In some problems of numerical analysis or optimization, we are faced with non-convergent sequences and have to obtain information from them (for examples see [23] , [24] , [26] , [30] and [37]). Sometimes we would like to know what kind of non-convergence it is, or how many accumulation points there are, or what the period is. In this chapter, we study this type of problem and try to determine what problems are decidable and what problems are not decidable.

In order to speak of decidability, we use the notion of algorithm in chapter 1, and reconsider the ideas of GOLD ([28] , [29]) : Let Q be a question and S be a family of sequences. We say that S is decidable in the limit on S if there exists an algorithm which for every sequence (x_n) in S gives a sequence of answers (r_n) which is correct when n is large enough. In the section 1, we give the definition and establish the theorem of normalization which states the equivalence of the algorithms for sequences and the normal algorithms on problems of decidability in the limit.

In section 2, we use the introduced notion on simple problems concerning convergence, periodicity and turbulence of sequences.

Section 3 is devoted to the problem of counting the number of accumulation points.

Section 4 is devoted to the problem of the determination in the limit of the period of an asymptotically periodic sequence. We define two families of algorithms, study them, and show that it is not possible to completely solve the problem.

Sequences obtained from an iteration $x_{n+1} = f(x_n)$ are a special case, and we study them separately in section 5. Two general results are established on the decidability in the limit and are given in section 6.

Three appendices complete the chapter. They are respectively devoted to the notion of strength of an accumulation point, to the possibility of defining the algorithms for sequences with conditions of recursivity and to the application of our methods to functional problems.

The results of this chapter have been previously published in [10]-, [15] and [19],[22].

Notation :

E   : metric space with distance d;

$B(x,\varepsilon)$: open ball with center y and radius ε

$$B(y,\varepsilon) = \{x \in E \mid d(x,y) < \varepsilon\}$$

$E^{(N)}$: the set of finite sequences of E :

$$E^{(N)} = \bigcup_{m \in N} E^m$$

E^N : the set of (infinite) sequences of E :

X^q_p : the section of points of $(x_n) \in E^N$ with indices
          i such that $p \le i \le q$:

$$X^q_p = (x_p , x_{p+1},\ldots,x_q) ;$$

$A(x_n)$: the set of accumulation points of the sequence (x_n) :

$$y \in A(x_n) <=> \forall \, \varepsilon > 0, \; \forall \, n \in N , \; \exists \, m \ge n$$
$$: x_m \in B(y,\varepsilon) ;$$

Alg(E,F) : the set of algorithms for sequences from E^N into
          F^N (see chapter 1) ;

Norm(E,F): the set of normal algorithms from E^N into F^N
          (see chapter 1);

Conv(E) : the family of convergent sequences of E ;

$Conv_0(E)$: the family of sequences of E which converge to 0
          (when E = **R** or R^m);

Stati(E) : the family of stationary sequences of points of E :

$$((x_n) \in Stati(E)) <=> (\forall \, n \in N \; x_{n+1} = x_n) ;$$

UStati(E): the family of ultimately stationary sequences of point of
          E :

$$((x_n) \in UStati(E)) <=> (\exists \, p \in N , \; \forall \, n \ge p :$$
$$x_{n+1} = x_n) ;$$

$Per_p(E)$: the family of periodic sequences with period p :

$$((x_n) \in Per_p(E)) <=> (p \text{ is the smallest positive}$$
          integer such that $\forall \, n \in N \; x_{n+p} = x_n)$

$$Per(E) = \bigcup_{p \in N^*} Per_p(E) ;$$

$Per^*_p(E) = \{(x_n) \in Per_p(E) | \forall\ i,j \in \{0,1,\ldots,p-1\}$
$\quad\quad\quad\quad i \neq j \Rightarrow x_i \neq xj\}$;

$Per^*(E) = \underset{p\in \mathbf{N}^*}{\cup}\ Per^*_p(E)$;

$UPer_p(E)$:   Family of ultimately periodic sequences with period p

$\quad\quad\quad\quad (x_n) \in UPer_p(E) <=>$
$\quad\quad\quad\quad (p$ is the smallest positive integer such that
$\quad\quad\quad\quad \exists\ n_0 \in \mathbf{N}\ ,\ \forall\ n \geq n_0 : x_n = x_{n+p})$;

$UPer_p(E) = \underset{p\in \mathbf{N}^*}{\cup}\ UPer_p(E)$;

$UPer^*_p(E) = \{(x_n) \in UPer_p(E) | \exists\ n_0 \in \mathbf{N}\ ,\ \forall\ n \geq n_0\ ,$
$\quad\quad\quad\quad \forall\ i,j \in \{n,n+1,\ldots,n+p-1\}$
$\quad\quad\quad\quad i \neq j \Rightarrow x_i \neq x_j\}$

$UPer^*(E) = \underset{p\in \mathbf{N}^*}{\cup}\ UPer^*_p(E)$;

$APer_p(E)$:   the family of asymptotically periodic sequences
$\quad\quad\quad\quad$ with period p :

$\quad\quad\quad\quad (x_n) \in APer_p(E) <=>$
$\quad\quad\quad\quad (p$ is the smallest positive integer such that
$\quad\quad\quad\quad$ sub sequences $(x_{np}),(x_{np+1}),\ldots,(x_{np+p-1})$
$\quad\quad\quad\quad$ are convergent) ;

$APer(E) = \underset{p\in \mathbf{N}^*}{\cup}\ APer_p(E)$;

$APer^*(E) = \underset{p\in \mathbf{N}^*}{\cup}\ APer^*_p(E)$;

$Fini_p(E) = \{(x_n) \in E^{\mathbf{N}} | card\ A(x_n) = p\}$;

$Fini(E) = \underset{p\in \mathbf{N}^*}{\cup}\ Fini_p(E)$;

$Turb(E) = \{(x_n) \in E^{\mathbf{N}} | A(x_n)$ is infinite$\}$.

$\rho \in \mathbf{R}^{+*}$

$Fini_\rho(E) = \{(x_n) \in Fini(E) | \forall\ x,y \in A(x_n)$

$\quad\quad\quad\quad x \neq y \Rightarrow d(x,y) \geq \rho\}$

$(\varepsilon(p)) \in Conv_0(\mathbf{R}^{+*})$

$Fini_{\varepsilon(p)}(E) = \{(x_n) \in Fini(E) | \exists\ n_0 \in \mathbf{N}\ \ V^-\ p \geq n_0$

$\quad\quad\quad\quad d(A(x_n),\ x_p) \leq \varepsilon(p)\}$.

1 – DEFINITION AND THE NORMALISATION THEOREM

The notion of decidability in the limit was introduced and studied by GOLD ([28],[29]). This notion is adapted here to our problems by the use of the algorithms defined in Chapter 1. This permits us to formulate positive results easily (always obtained by explicit algorithms) and negative results which indicate the bounds of algorithmic possibilities concerning sequence transformations. The definitions and vocabulary are the same as in [19] .

Definition

Let Q be a question defined on $S \subset E^N$ (for example "does the sequence (x_n) converge ?" , "how many accumulation points does the sequence (x_n) have ?" . Let R be the set of possible answers [1] (for example : {YES,NO} , {0,1,...,n,...}). Let T be a sequence transformation from E^N into R^N . We say that T is **satisfactory for Q on S** if, for every sequence $(x_n) \in S$, the sequence of answers given by T is correct for n sufficiently large. We say that **Q is decidable in the limit** on S if there exists an algorithmic transformation ($T \in {}^T Alg(E,R)$) satisfactory for Q on S . Conversely, if this is not so, we say that Q is **undecidable in the limit** on S.

Remarks

1) When R = {YES,NO} we could use "predicate" instead of "question", as in logic. But, because we wish to consider bigger R , we prefer to use the word "question".

2) Since simple decidability may not expected concerning our problems, we will only deal with decidability in the limit. When R = {YES,NO} we can envisage "semi-decidability" : Q is semi-decidable in the limit on S if there exists an algorithmic transformation such that for every sequence $(x_n) \in S$
    [the sequence of answers is YES for n large enough]

$$\langle = \rangle$$

    [the correct answer is YES]

    It is easy to show that Q is decidable in the limit on S if and only if Q and not-Q are semi-decidable in the limit on S . We shall go no further with semi-decidability in the limit.

----

[1] on R we consider the distance d defined by d(x,y) = 0 if x = y , d(x,y) = 1 if $x \neq y$.

3) We define decidability in the limit without prescribing calculability conditions. If doing so, we follow the usage of the
   word algorithm in numerical analysis, particularly in optimization theory. But this point of view is not restrictive :

> positive results are established by giving explicit
> algorithms which are always calculable (when it is possible
> to give a meaning to the word calculable).

> negative results are a fortiori true if we prescribe
> calculability conditions (see appendix 2).

Our point of view also has the advantage of giving a sense
to the notion of decidability in more cases than the usual ones
(for example, we are not obliged to consider only denumerable
spaces E and R); this point of view permit us to do not
deal with delicate notions (not really necessary).

Furthermore, this point of view emphasizes the nature of the
impossibility of certain questions, which is due not to the
limited number of calculable functions, but to the lock of
information given to our algorithms at any stage of the calculation.

Theorem 1 (normalization theorem)

There exists a normal transformation satisfactory for Q on **S** if
and only if :

> there exists an algorithmic transformation satisfactory for
> Q on **S** .

Proof

$\Rightarrow$ Obvious since $^\tau \text{Norm}(E,R) \subset {}^\tau \text{Alg}(E,R)$

$\Leftarrow$ Let $A = (R,S) \in \text{Alg}(E,R)$ be satisfactory for Q on $S \subset E^N$
   and let $r \in R$.

We shall define a normal algorithm $N = (f_n)$ which will be satisfactory for Q on **S**.

For every sequence $(x_n) \in S$, we define a sequence (n_j) as follows
: "n_i is the biggest index of points used in the calculation of

$$A^{(0)}(x_n) , A^{(1)}(x_n),\ldots,A^{(i)}(x_n)."$$

For every j we set :

$$f_j(x_0,x_1,\ldots,x_j) = r \quad \text{if} \quad j < n_0 ;$$

$f_j(x_0, x_1, \ldots, x_j) = A^{(i)}(x_n)$ if $j \geq n_0$ and if there exists

i such that $n_i \leq j < n_{i+1}$;

$f_j(x_0, x_1, \ldots, x_j) = A^{(j)}(x_n)$ if $j \geq n_0$ anf if, for every i

$j > n_i$

The functions f_j are well defined (i.e. the definition of

$f_j(x_0, x_1, \ldots, x_j)$ depends only on j and

$(x_0, x_1, \ldots, x_j)$

and does not depend on the rest of (x_n)). The sequence given by N
is the same as the one given by A except for r terms in the beginning and repetition. Hence N is satisfactory for Q on S.

Remark

This theorem means that for all problems of decidability in the limit,
what we can do with general algorithms (i.e. in **Alg**(E,R)) can also
be done with normal algorithms.

But it is easy to see that the question "is (x_n) stationary ?" is
decidable in the limit on $E^{\mathbf{N}}$ by normal algorithms and is not decidable in the limit by algorithms with k-memories (for any k) , if
card $E \geq 2$: algorithms with k-memories solve strictly fewer problems
than normal algorithms.

We can easily find questions :

- decidable in the limit by algorithms with (k+1) memories and
  undecidable in the limit by algorithms with k-memories.

- decidable in the limit by algorithms with k-memories and
  undecidable in the limit by k'-stationary algorithms.

- decidable in the limit by (k+1)-stationary algorithms and
  undecidable in the limit by k-stationary algorithms.

2 - PROBLEMS CONCERNING CONVERGENCE, TURBULENCE AND PERIODICITY
        OF SEQUENCES

Since at each step (i.e. for each answer), an algorithmic sequence
transformation uses only a finite set of points of (x_n), it is clear
that such transformation cannot be satisfactory about question such
"is (x_n) a convergent sequence ?" ; some global information is
necessary. To give such information means to say $(x_n) \in S$ where S
is a family of sequences; if S is small, this information is precise

and thus it becomes easy to find an algorithmic transformation satisfactory for Q on **S** .

When Q is fixed, to find what additional information is necessary is equivalent to finding for what family $S \subset E^N$ there exists an algorithmic transformation satisfactory for Q on **S**. This is what we do here.

Results obtained on convergence, periodicity and turbulence show that rather precise additional information is often required.

This work began in $[16]$ and has continued in $[19]$ (from which we use the presentation) and $[14]$.

Theorem 2

The question "is (x_n) a convergent sequence ?" is decidable in the limit.

    (i) on **UPer**(E) ;
    (ii) on $\mathbf{APer}_\varepsilon(E)$ for every $\varepsilon > 0$;

and is undecidable in the limit

    (j)   on **APer**(E) , if $E^\alpha \neq \emptyset$;
    (jj) on **UStati**(E) $\cup$ **Turb**(E), if E is infinite;
  (jjj) on $\mathbf{Fini}_1(E)$, if E is not compact;
    (jw) on **UStati**(E) $\cup$ $\mathbf{Fini}_p(E)$ for every $p > 1$,
          if card (E) $\geq$ p.

E^α is the set of accumulation points of E;

$\mathbf{APer}_\varepsilon(E) = \{(x_n) \in \mathbf{APer}(E) | \; \forall \; x,y \in A(x_n) : x \neq y \Rightarrow d(x,y) \geq \varepsilon\}.$

Remarks

1)   Result (i)   points out that to know that (x_n) is ultimately periodic allows us to decide its convergence. Result (j)   points out that to know that (x_n) is asymptotically   periodic is not sufficient.   Result (ii)   means that if one knows that (x_n) is asymptotically periodic and its accumulation points are mutually distant at less than ε (ε fixed) then again, one can decide its convergence. Result (jj)   shows that it is impossible to distinguish in the limit between an ultimately stationary sequence and a turbulent one.

2)   Of course, if a family **S** is contained in a family for which the question of convergence is decidable in the limit, then the same holds for **S**. Conversely, if **S** contains a family for which the question of convergence is undecidable in the limit, then the same holds for **S**. This remark, with theorem 2 and lemmas (sometimes more precise) allows us to answer the question of decidability in the limit of convergence for all the families studied here.

3) The conditions concerning E given in (j) (jj) (jjj) ans (jw)
in fact are necessary and sufficient.
Indeed :

* If $E^{\alpha} = \emptyset$ then **APer**(E) = **UPer**(E) , and thus
  (from (i)), the question is decidable on **APer**(E) ;

* If E is finite, **UStati**(E) ∪ **Turb**(E) = **Conv**(E) ;

* If E is compact : $\mathbf{Fini}_1(E) = \mathbf{Conv}(E)$;

* If card(E) < p , then **UStati**(E) ∪ $\mathbf{Fini}_p(E)$ = **UStati**(E).

Proof

(i) Result(i) follows immediately from the next lemma.

Lemma 1

The question "is (x_n) a convergent sequence ?" is decidable in the
limit on :

   UStati(E) ∪ (**APer**(E) − **Conv**(E)).

Proof of lemma 1

We define a normal algorithm by :

   $f_i(x_0, x_1, \ldots, x_i) = \text{YES} \Longleftrightarrow (x_{[i/2]} = x_{[i/2]+1} = \cdots = x_i)$

($[a]$ is the integer part of a)

This algorithm is satisfactory for the question of convergence.
Indeed, let $(x_n) \in \mathbf{UStati}(E) \cup (\mathbf{APer}(E) - \mathbf{Conv}(E))$. Two cases are
possible :

−a− (x_n) is ultimately stationary, so there exists $i_0 \in \mathbf{N}$
     such that if $i \geq i_0$ then :

$$x_{[i/2]} = x_{[i/2]+1} = \cdots = x_i \; ,$$

   and thus the algorithm is satisfactory for (x_n).

−b− (x_n) is asymptotically periodic with period $p \geq 2$,
     then there exists j_0 such that :

$$\forall \, n \in \mathbf{N} \quad x_{j_0 + np} \neq x_{j_0 + np + 1}$$

   Thus for every $i > \max(2p, 2j_0)$:

$$f_i(x_0, x_1, \ldots, x_i) = \text{NO} \; .$$

Hence the algorithm is satisfactory for (x_n)

(ii) This is a consequence of proposition 2 of section 2.

(j) Result(j) follows immediately from the next lemma.

Lemma 2

Let E statisfy $E^\alpha \neq \emptyset$.

The question "is (x_n) a convergent sequence ?"
is undecidable in the limit on $\text{Conv}(E) \cup \text{UPer}_p(E)$ for every
$p \in \mathbf{N}$, $p \geq 2$.

Proof of the lemma 2

Assume that (f_n) is a normal algorithm satisfactory for convergence
on $\text{Conv}(E) \cup \text{UPer}_p(E)$, with $p \geq 2$ given.

Let (a_i) be a sequence of points of E which converges to $a \in E^\alpha$
and such that :

$$\forall n \in \mathbf{N} \quad a_i \neq a ,$$

$$\forall i,j \in \mathbf{N} \quad i \neq j \Rightarrow a_i \neq a_j .$$

For every $i \in \mathbf{N}$, we define a sequence (y^i_n) by :

$$y^i_n = a_i \quad \text{if } i \text{ is even and } p \text{ divides } n ,$$

$$y^i_n = a \quad \text{if not.}$$

Set $(x^0_n) = (y^0_n)$. This sequence is periodic with period p , thus
there exists $n_0 \in \mathbf{N}$ such that $f_n(x^0{}_0, x^0{}_1, \ldots, x^0{}_{n_0}) = \text{NO}$

Set $x^1_n = x^0_n$ if $n \leq n_0$,

$x^1_n = y^1_n$ if $n > n_0$.

This sequence is convergent. Thus there exists $n_1 > n_0$ such that

$$f_{n_1}(x^1{}_0, x^1{}_1, \ldots, x^1{}_{n_1}) = \text{YES}$$

Set $x^2_n = x^1_n$ if $n \leq n_1$

$x^2_n = y^2_n$ if $n > 1$

The sequence (x^2_n) is periodic with period p . Thus there exists
$n_2 > n_1$ such that

$$f_{n_2}(x^2{}_0, x^2{}_1, \ldots, x^2{}_{n_2}) = \text{NO}$$

Continuing, we obtain sequences

$$(x^0{}_n),(x^1{}_n),(x^2{}_n),\ldots,(x^i{}_n),\ldots,$$

and $n_0 < n_1 < n_2 < \ldots < n_i < \ldots$ such that :

if i is even $f_{n_i}(x^i{}_0,x^i{}_1,\ldots,x^i{}_{n_i}) =$ NO

if i is odd $f_{n_i}(x^i{}_0,x^i{}_1,\ldots,x^i{}_{n_i}) =$ YES

Hence the sequence :

$$(x_n) = (x^0{}_0,x^1{}_0,\ldots,x^0{}_{n_0},x^1{}_{n_0+1},\ldots,x^1{}_{n_1},x^2{}_{n_1+1},\ldots)$$

satisfies :

if i is even, $f_{n_i}(x_0,x_1,\ldots,x_{n_i}) =$ NO,

if i is odd, $f_{n_i}(x_0,x_1,\ldots,x_{n_i}) =$ YES.

By construction, the sequence (x_n) converges, and consequently the algorithm (f_n) is not satisfactory on Conv(E) ∪ UPer$_p$(E).

Remarks

1) The proof of Lemma 2 in fact shows that the question "is (x_n) a convergent sequence ?" is not semi-decidable in the limit on Conv(E) ∪ UPer$_p$(E). An additionnal study shows that the question "is (x_n) a non convergent sequence ?" is semi-decidable in the limit on Conv(E) ∪ UPer$_p$(E) (even on Conv(E) ∪ UPer(E)).

2) Easy modifications of the proof of lemma 2 show that the convergence is undecidable in the limit on Conv(E) ∪ UPer$^*{}_p$(E). Thus in theorem (j) , one can replace APer(E) by APer*(E).

Continuation of the proof of theorem 2

(jj) Let $(y^0{}_n)$, $(y^1{}_n),\ldots,(y^i{}_n),\ldots$ be sequences defined by :
let (β_i) a sequence of distinct points of E
$((\beta_i)$ exists since E is infinite).

We set :

$y^i{}_n = \beta_{h(n)}$ if i is even

$y^i{}_n = \beta_0$ if not,

where $(h(n)) = (0,0,1,0,1,2,0,1,2,3,0,1,2,3,4,0,\ldots)$

From sequences (y^i_n) , we proceed as in lemma 2 , but with the following condition :

$$n_i > 2\, n_{i-1} .$$

If i is even, then (x^i_n) is turbulent, and if i is odd, then (x^i_n) is ultimately stationary. From the condition $n_i > 2\, n_{i-1}$, we obtain that the final sequence (x_n) is turbulent. By construction, for every odd i we have :

$$f_{n_i} (x_0,x_1,\ldots,x_{n_i}) = YES$$

(jjj) Let $(y^0_n),(y^1_n),\ldots,(y^i_n),\ldots$ be sequences defined by : let (γ_i) be a sequence with no accumulation point. We set :

$$y^i_n = \gamma_n \text{ if } i \text{ is even and } n \text{ is even,}$$

$$y^i_n = \gamma_0 \text{ if not.}$$

We proceed as in (j) .

(jw) Let $(y^0_n),(y^1_n),\ldots,(y^i_n),\ldots$ be sequences defined by : let $\delta_1,\delta_2,\ldots,\delta_p$, p distinct points of E . We set :

$$yn = \delta_{\overline{n}^p} \text{ if } i \text{ is even,}$$

$$y^i_n = \delta_0 \text{ if not,}$$

where $\overline{n}^p = n-p . [n/p]$

We work as in (jj) , with the condition that $n_i > n_{i-1} + p$.

Theorem 3

Consider the questions Q : "is (x_n) periodic ?"
                        Q' : "is (x_n) ultimately periodic ?"
                        Q" : "is (x_n) asymptotically periodic ?"

       (i)   Q is decidable in the limit on :

$$\bigcup_{p=1}^{n} UPer_p(E) \cup Turb(E) \text{ for every } n \in \mathbb{N}^* ;$$

  (ii)   Q' and Q" are decidable in the limit on

$$\bigcup_{p=1}^{n} Per_p(E) \cup Turb(E) \text{ for every integer } n ;$$

(j)   Q is undecidable in the limit on :

Per(E) ∪ Turb(E) , if E is infinite;

(jj) Q' and Q" are undecidable in the limit on :

$$\bigcup_{p=1}^{n} UPer_p(E) \cup Turb(E) \quad \text{for every} \quad n \geq 2 ,$$

if E is infinite;

(jjj) Q , Q' and Q" are undecidable in the limit on :

Fini(E) , if E has more than one element.

Sketch of the proof of theorem 3

(i) (ii)

$$f_i(x_0, x_1, \ldots, x_i) = \text{YES} \iff \left| \begin{array}{l} \text{the sequence } (x_0, x_1, \ldots, x_i) \\ \text{is periodic of period } p \leq n \end{array} \right|$$

(j)

We proceed as in proof of (j) (jj) (jjj) and (jw) of theorem 2 using the following sequences :

$(x^0_n) = (\beta_0, \beta_0, \beta_1, \beta_0, \beta_1, \beta_2, \ldots, \beta_0, \beta_1, \beta_2, \ldots, \beta_i, \beta_0, , , ,)$;

$(x^1_n) = (x^0_0, x^0_1, \ldots, x^0_{n_0}, x^0_0, x^0_1, \ldots, x^0_{n_0}, x^0_0, x^0_1, \ldots, x^0_{n_0}, \ldots)$;

$(x^2_n) = (x^1_0, x^1_1, \ldots, x^1_n, \beta_0, \beta_0, \beta_1, \beta_0, \beta_1, \beta_2, \ldots)$;

$(x^3_n) = (x^2_0, x^2_1, \ldots, x^2_{n_2}, x^2_0, x^2_1, \ldots, x^2_{n_2}, x^2_0, x^2_1, \ldots, x^2_{n_2}, \ldots)$;

etc...

(jj)

$(x^0_n) = (\beta_0, \beta_0, \beta_1, \beta_0, \beta_1, \beta_2, \ldots, \beta_0, \beta_1, \beta_2, \ldots, \beta_i, \beta_0, \ldots)$;

$(x^1_n) = (x^0_0, x^0_1, \ldots, x^0_{n_0}, \beta_0, \beta_1, \beta_0, \beta_1, \beta_0, \beta_1, \ldots)$;

$(x^2_n) = (x^1_0, x^1_1, \ldots, x^1_{n_1}, \beta_0, \beta_0, \beta_1, \beta_0, \beta_1, \beta_2, \ldots)$;

$(x^3_n) = (x^2_0, x^2_1, \ldots, x^2_n, \beta_0, \beta_1, \beta_0, \beta_1, \beta_0, \beta_1, \ldots)$

etc...

(jjj)

$(x^0_n) = (\alpha, \beta, \alpha, \alpha, \beta, \alpha, \alpha, \alpha, \beta, \alpha, \alpha, \alpha, \alpha, \beta, \ldots)$;

$$(x^1{}_n) = (x^0{}_0, x^0{}_1, \ldots, x^0{}_{n_0}, x^0{}_0, x^0{}_1, \ldots, x^0{}_{n_0}, x^0{}_0, x^0{}_1, \ldots, x^0{}_{n_0}, \ldots) \ ;$$

$$(x^2{}_n) = (x^1{}_0, x^1{}_1, \ldots, x^1{}_{n_1}, \alpha, \beta, \alpha, \alpha, \beta, \alpha, \alpha, \alpha, \beta, \ldots) \ ;$$

$$(x^3{}_n) = (x^2{}_0, x^2{}_1, \ldots, x^2{}_{n_2}, x^2{}_0, x^2{}_1, \ldots, x^2{}_{n_2}, \ldots, x^2{}_0, x^2{}_1, \ldots$$

$$\ldots, x^2{}_{n_2}, \ldots, x^2{}_0, x^2{}_1, \ldots, x^2{}_{n_2}, \ldots) \ ;$$

etc...

Remarks

1) Theorem 3 seems rather disappointing, for the families of sequences for which Q, Q' and Q" are decidable in the limit, are small families and the algorithm which gives the decision is almost obvious. However, part (j), (jj) and (jjj) of the theorem show that it is no use to look for more elaborate results, for if one enlarges the family of (i), Q becomes undecidable and if one enlarges the family of (ii), Q' and Q" become undecidable.

2) The conditions given in (j), (jj) and (jjj) are necessary and sufficient.

3) The sketch of the proof for (j), (jj) and (jjj) allows us to show that on the three families considered, Q , Q' and Q" are not semi-decidable in the limit, and the same holds for their negation.

Theorem 4

The question "is (x_n) a turbulent sequence ?" is decidable in the limit on :

(i) $\displaystyle\bigcup_{p=1}^{n} \mathrm{Per}_p(E) \cup \mathrm{Turb}(E)$ for every integer n;

and is undecidable in the limit on :

(j) $\mathrm{UStati}(E) \cup \mathrm{Turb}(E)$, if E is infinite;

(jj) $\mathrm{Per}(E) \cup \mathrm{Turb}(E)$, if (E) is infinite.

Proof

(i) follows theorem 3 (ii)
(j) follows theorem 2 (jj)
(jj) follows theorem 3 (j).

3 — ALGORITHMS FOR COUNTING THE NUMBER OF ACCUMULATION POINTS

In this section, we present two algorithms for counting the number of
accumulation points of sequences in certain explicitly given families.
The domains of efficiency of these algorithms are distinct and to
define them, we need the notion of strength of an accumulation point
and of the quickness of a sequence (see appendix 1). These two algorithms are simple and natural. With the negative results concerning
this type of problem, we can conclude by saying that to count the
number of accumulation points, whether the algorithms exist and are
simple and natural, or they don't exist. This type of problem was
first considered in [10], [11] and [13].

Let   E   be a metric space with distance   d .

Let $\varepsilon(p) \in \text{Conv}_0(\mathbf{R}^+)$ and let $(\beta(p))$ be a sequence of integers such
that

$$\forall \, p \in \mathbf{N} \, , \, \beta(p) > p \, .$$

Algorithm $\text{NAP}[\varepsilon(p), \beta(p)]$

Step p

Points $x_p, x_{p+1}, \ldots, x_{\beta(p)}$ can be used.

The proposed answer is r_p ,   the number of connected components of
the (non-oriented) graph defined by vertices $x_p, x_{p+1}, \ldots, x_{\beta p}$ and
edges $\{x_i, x_j\}$ such that :

$$d(x_i, x_j) \leq \varepsilon(i) + \varepsilon(j) \, .$$

Example

$$E = \mathbf{R} \, ; \, \varepsilon(p) = \frac{1}{p+1} \, ; \, \beta(p) = 2p \, ;$$

$$(x_n) = (1, 2 + \frac{1}{2} \, , \, 4 + \frac{1}{3} \, , \, \frac{1}{4} \, , \, 2 + \frac{1}{5} \, , \, 4 + \frac{1}{6} \, , \, 2 + \frac{1}{8} \, , \ldots)$$

Step 0

The graph has one vertex x_0 and no edge : $r_0 = 1$

Step 1

The graph has two vertices x_1 , x_2 and no edge : $r_1 = 2$

Step 2

The graph has 3 vertices x_2, x_3, x_4 and no edge : $r_2 = 3$

Step 3

The graph has 4 vertices x_3, x_4, x_5, x_6 and one edge $\{x_3, x_6\}$: $r_3 = 3$

Step 4

The graph has 5 vertices x_4, x_5, x_6, x_7, x_8 and two edges :
$\{x_4, x_7\}$, $\{x_5, x_8\}$ $r_4 = 3$

etc...

We verify that for all $p \geq 3$: $r_p = 3$. The choice of the sequence $\beta(p)$ is essential. If the accumulation points of (x_n) are difficult to detect (for example : point of strength zero), we have to choose $\beta(p)$ to be rapidly increasing. Particularly :

Proposition 1

If $\beta(p)$ satisfies : $\exists\, p_0 \in \mathbf{N}$, $\forall\, p \geq p_0$: $\beta(p) \geq p^2$
and if $\varepsilon(p) \in \mathrm{Conv}_0(\mathbf{R}^+)$,
then the algorithm $\mathrm{NAP}[\varepsilon(p), \beta(p)]$ is satisfactory for the question "what is the number of accumulation point of (x_n) ?" on $\mathrm{Fini}^+{}_{\varepsilon(p)}(E)$.

$\mathrm{Fini}^+{}_{\varepsilon(p)}(E)$ is the sequence family defined by :

$(x_n) \in \mathrm{Fini}^+{}_{\varepsilon(p)}(E)$ $<=>$ $\begin{vmatrix} (x_n) \text{ has a finite number of accumulation} \\ \text{points,} \\ \text{each of them has a strictly positive} \\ \text{strength and is of quickness } \varepsilon(p) \\ (\text{see appendix 1}) \end{vmatrix}$

Proof

Let $(x_n) \in \mathrm{Fini}^+{}_{\varepsilon(p)}$ having accumulation points $y_1, y_2, \ldots, y_k$, and let $\varepsilon > 0$ be defined by :

$$\varepsilon = \min\{d(y_i, y_j) \,|\, i, j \in \{1, 2, \ldots, j\}\ i \neq j\} .$$

Let p_0 be such that for every $p \geq p_0$:

$\begin{vmatrix} d(x_p, \{y_1, y_2, \ldots, y_k\}) \leq \varepsilon(p) \leq \varepsilon/5 , \\ \forall\, \ell \in \{1, 2, \ldots, k\} : B(y_\ell, \varepsilon/5) \cap X^{\beta(p)}{}_p \neq \emptyset . \end{vmatrix}$

We shall prove that, for every $p \geq p_0$, $r_p = k$.

It is easy to see that :

(a) $\forall\, \ell_1$, $\ell_2 \in \{1, 2, \ldots, k\}$: $\ell_1 \neq \ell_2 =>$

$d(B(y_{\ell_1}, \varepsilon/5) \cap X_p^{\beta(p)}, B(y_{\ell_2}, \varepsilon/5) \cap X_p^{\beta(p)}) > 2\,\varepsilon/5$;

(b) $\forall\ n_1, n_2 \in \{p, p+1, \ldots, \beta(p)\}$; $n_1 \neq n_2$,

$$\left.\not\exists\ \ell\ \begin{array}{c} x_{n_1} \in B(y_\ell, \varepsilon/5) \\[10pt] x_{n_2} \in B(y_\ell, \varepsilon/5) \end{array}\right| \Rightarrow d(x_{n_1}, x_{n_2}) \leq \varepsilon(n_1) + \varepsilon(n_2)$$

From (a), we deduce that the number of connected components is at least k . From (b), we obtain that it is at most k .

Let $\rho \in \mathbf{R}^+$. Let $\alpha(p)$, $\beta(p)$ be two sequences of integers such that :

$\forall\ p \in \mathbf{N}$: $\beta(p) \geq \alpha(p) \geq p$.

Algorithm NAP'$\{\rho, \alpha(p)\ ,\ \beta(p)\}$

Step p.

Points $(x_{\alpha(p)}, x_{\alpha(p)+1}, \ldots, x_{\beta(p)})$ can be used.

The proposed answer is r_p , the number of connected components of the (non-oriented) graph defined by vertices $x_{\alpha(p)}, x_{\alpha(p)+1}, \ldots, x_{\beta(p)}$ and edges $\{x_i, x_j\}$ such that :

$$d(x_i, x_j) \leq 2\rho/5\ .$$

Example

$E = \mathbf{R}$; $\rho = 1$; $\alpha(p) = 5\ p$; $\beta(p) = 5p+4$;

$(y_n) = (3, 2, 3, 1, 2, 3, 2, 3, 1, 2, 3, 2, 3, 1, 2, \ldots)$;

$(z_n) = (2, -2, 1, -1, 1/2, -1/2, 1/4, -1/4, 1/8, -1/8, \ldots)$;

$(x_n) = (y_n + z_n)$.

Step 0

The graph has 5 vertices :

$x_0 = 5$; $x_1 = 0$; $x_2 = 4$; $x_3 = 0$; $x_4 = 2.5$,

and there is one edge : $\{x_1, x_3\}$; thus $r_0 = 4$.

Step 1

The graph has 5 vertices :

$x_5 = 2.5$; $x_6 = 2.25$; $x_7 = 2.75$; $x_8 = 1.125$; $x_9 = 1.875$

and there are 3 edges : $\{x_5, x_6\}$, $\{x_5, x_7\}$, $\{x_9, x_6\}$; thus $r_1 = 2$.

Step 2

The graph has 5 vertices :

$$x_{10} = 3.0625 \; ; \; x_{11} = 1.9375 \; ; \; x_{12} = 3.03125 \; ;$$
$$x_{13} = 0.96875 \; ; \; x_{14} = 2.015625$$

and there are 2 edges : $\{x_{11}, x_{14}\}$, $\{x_{10}, x_{12}\}$; thus $r_2 = 3$, etc...

It is easy to verify that, for $p \geq 2$, there are 5 vertices and 2 edges; so that $r_p = 3$.

Proposition 2

Let $\rho \in R^{+*}$. Let E be a compact subset of R^m. The algorithm NAP'$[\rho, p, p^2]$ is satisfactory for the question "what is the number of accumulation points of (x_n) ?" on $\text{Fini}_\rho^+(E)$.

$\text{Fini}_\rho^+(E)$ is the family of sequences defined by :

$$(x_n) \in \text{Fini}_\rho^+(E) \Leftrightarrow \left| \begin{array}{l} (x_n) \text{ has a finite number of accumulation} \\ \text{points} \\ y_1, \ldots, y_\ell \text{ , each of them has a strictly} \\ \text{positive strength and } \forall \; i, j \in \{1, \ldots, \ell\} : \\ i \neq j \Rightarrow d(y_i, y_j) \geq \rho. \end{array} \right.$$

Proof

Let $(x_n) \in \text{Fini}_\rho^+(E)$ and let $y_1, y_2, \ldots, y_\ell$ be its accumulations points..

We set :

$$V = B(y_1, \rho/5) \cup \ldots \cup B(y_\ell, \rho/5).$$

There exists p_0 such that : $\forall \; p \geq p_0 : x_p \in V$. (*).

(p_0 exists because E is assumed to be a compact subset of R^m) .

There exists $p_1 \geq p_0$ such that for every $i \in \{1, 2, \ldots, \ell\}$:

$$\forall \; p \geq p_1 : X_p^{p^2} \cap B(y_i, \rho/5) \neq \emptyset \quad (**) .$$

Let $p \geq p_1$; from (*) , we have :

$$\forall \; \ell_1, \ell_2 \in \{1, 2, \ldots, \ell\} : \ell_1 \neq \ell_2 \Rightarrow$$

$$d(B(y_{\ell_1}, \rho/5) \cap X_p^{p^2} , B(y_{\ell_2}, \rho/5) \cap X_p^{p^2}) \geq 2 \, \rho/5 ,$$

With (**), we obtain $r_p \geq \ell$.

The relation (*) gives :

$$\forall\ n_1, n_2 \in \{p, p+1, \ldots, p^2\}\ ,\ n_1 \neq n_2\ ,$$

$$\left. \begin{array}{l} x_{n_1} \in B(y_{\ell'}, \rho/5) \\ \\ \not\exists\ \ell' \\ \\ x_{n_2} \in B(y_{\ell'}, \rho/5) \end{array} \right| \quad \Rightarrow\ d(x_{n_1}, x_{n_2}) \le 2\ \rho/5$$

Thus $r_p \le \ell$. Finally , $r_p = \ell$.

Remark.

The coefficient $2\rho/5$ in the definition of NAP' may be replaced by $h\rho$ where $o < h < 1/2$.

Theorem 5

The question "What is the number of accumulation points of (x_n) ?" is decidable in the limit :

(i)   on $\text{Fini}^+_{\epsilon(p)}(E)$ (for every $\epsilon(p) \in \text{Conv}_0(R^+)$)

(ii)   on $\text{Fini}^+_\rho(E)$ (for every $\rho \in R^{+*}$ and compact E)

and is undecidable in the limit

(j)   on $\text{APer}^*(E)$ (if $e^{\alpha\alpha} \neq \emptyset$)

Proof

(i)   proposition 1

(ii)   proposition 2

(j)   proposition 5 of § 4.

4 —   ALGORITHM FOR DETERMINING THE PERIOD OF AN ASYMPTOTICALLY PERIODIC SEQUENCE.

The problem of the algorithmic determination of the period of a sequence is interesting, for there is no global solution (proposition 5), and various approaches are possible, each one giving rise to an efficient algorithm with a different domain of efficiency.

The first method (the method of detector coefficients), is based on the calculation of means of mutual distance between points of the sequence; the study of these means allows to find (or to presume) the period of the considered sequence. Depending on the nature of the global information we have, four algorithms are possible, each having its specific set of efficiency (see the numerical experiments). The second method (method of barycenters) is based on the idea that when a sequence is asymptotically periodic, all the points of the sequence are close to certain barycenters of points of the sequence.

These algorithms need more computation than those obtained with the first method, but they are efficient (as seen by the numerical experiments).

This section is a synthesis of the following publications : [10],[11],[12],[13],[14],[19] and [20].

Notation

If k and n are two integers ($k \geq 1$), we denote by $\bar{n}^k$ (or $\bar{n}$ if k is given without ambiguity), the remainder of the euclidian division of n by k ($\overline{20}^7 = 6$).

a) Method of detector coefficients

Let (x_n) be a sequence of points of the metric space E .

We set :

$$i_1(p) = \sum_{i=1}^{2p-1} d(x_i, x_{i+1})/(2p-1) \; ;$$

$$i_2(p) = \sum_{i=1}^{2p-2} d(x_i, x_{i+2})/(2p-2) \; ;$$

$$\cdots$$

$$i_p(p) = \sum_{i=1}^{p} d(x_i, x_{i+p})/p \; .$$

Lemma 1

If $(x_n) \in APer^*_k(E)$ then :

(i) for every $r \in \mathbf{N}^*$, $i_r(p)$ converges;

If r is a multiple of k ,
then $\lim_{p \to \infty} i_r(p) = 0$.

If r is not a multiple of k then

$$\lim_{p \to \infty} i_r(p) \geq \varepsilon_0 = \min\{d(y_j, y_{j'}) \mid j \neq j'\} > 0$$

$$\leq \varepsilon_1 = \max\{d(y_j, y_{j'}) \mid j \neq j'\}$$

$(y_0, y_1, \ldots, y_{k-1}$ are the limits of
$$(x_{nk}), (x_{nk-1}), \ldots, (x_{nk+k-1}))$$

(ii)   there exists $p' \in \mathbf{N}$ such that, for every $p \geq p'$:

(P_k)
$$\left. \begin{array}{l} \forall\ j' \in \{2,3,\ldots,p\}\ k \mid j' \\[2mm] \forall\ j'' \in \{2,3,\ldots,p\}\ k \nmid j'' \end{array} \right| \quad \Rightarrow\ i_{j'}(p) \leq i_{j''}(p)$$

$(k \mid j'$ means k divides j'' ;
$k \nmid j''$ means k does not divide $j'')$.

Proof

(i)   Let r be a multiple of k ;   the sequence
    $(d(x_i, x_{i+r}))_i$ converges to 0 , hence the sequence :

$$\left(\sum_{i=1}^{2p-1} d(x_i, x_{i+r})/(2p-r) \right)_p$$

converges to 0 (this is a particular case of the
well-known result :

$$\lim_{n \to \infty} V_n = \ell \Rightarrow \lim_{n \to \infty} \sum_{i=1}^{n} V_i/n = \ell)$$

Now let r be a non-multiple of k . The following
sequences :

$(d(x_{nk}, x_{nk+r}))_n$

$(d(x_{nk+1}, x_{nk+1+r}))_n$

$\cdots \quad \cdots$

$(d(x_{nk+k-1}, x_{nk+k-1+r}))_n$

converge respectively to

$d(y_0, \overline{y_r})$, $d(y_0, \overline{y_{r+1}})$, $\ldots$, $d(y_{k-1}, \overline{y_{r+k-1}})$, thus the sequence :

$$\left(\left(\sum_{\substack{n \\ 1 \le nk \le 2p-r}} d(x_{nk}, x_{nk+r})/(2p-r)\right)_p =\right.$$

$$\left(\sum_{\substack{n \\ 1 \le nk \le 2p-r}} \frac{1}{k} \; d(x_{nk}, x_{nk+r})/(\frac{2p-r}{k})\right)_p$$

converges to $d(y_0, \overline{y_r})/k$.

Similarly :

$$\left(\sum_{\substack{n \\ 1 \le nk+1 \le 2p-r}} d(x_{nk+1}, x_{nk+1+r})/(2p-r)\right)_p$$

converges to $d(y_1, \overline{y_{1+r}})/k$,

etc...

Summing, we obtain :

$$\lim_{p \to \infty} i_r(p) = \sum_{i=0}^{k-1} d(y_i, \overline{y_{i+r}})/k \begin{array}{l} \ge \varepsilon_0 \\ \le \varepsilon_1 \end{array}.$$

(ii) Let p_0 be such that :

$$\forall \; p \ge p_0 : d(x_p, \overline{x_p}) \le$$

Let $p_1 \ge p_0$ be such that :

$$\forall \; p \ge p_1 : \sum_{j=0}^{k-1} \sum_{i=1}^{p_0} d(x_i, x_j)/p \le \varepsilon' = \varepsilon_0/16$$

and $\displaystyle\sum_{i<j \le p_0} d(x_i, x_j)/p \le \varepsilon'' = \varepsilon_0/16$.

Let $p' \ge 4p_1$. If $p \ge p'$ and $r \in \{2,3,...p\}$
r not a multiple of k , we have :

$$d(x_i, x_{i+r}) \ge d(\overline{y_i}, x_{i+r}) - d(x_i, \overline{y_i})$$

$$\ge d(\overline{y_i}, \overline{y_{i+r}}) - d(\overline{y_{i+r}}, x_{i+r}) - d(x_i, \overline{y_i})$$

$$\ge \varepsilon_0 - 2\varepsilon = 3 \; \varepsilon_0/4 .$$

Hence :

$$i_r(p) = \frac{\sum\limits_{i=1}^{2p-1} d(x_i,x_{i+r})}{2p-r} \geq \frac{\sum\limits_{i=p_0}^{2p-r} d(x_i,x_{i+r})}{2p-r}$$

$$\geq \frac{2p-r-p_0}{2p-r} \; \frac{3}{4} \; \varepsilon_0 \geq (1 - \frac{p_0}{p}) \; \frac{3}{4} \; \varepsilon_0$$

$$\geq \frac{3}{4} \cdot \frac{3}{4} \; \varepsilon_0 \geq \frac{\varepsilon_0}{2} \qquad (*)$$

Now, let r be a multiple of k . We have :

$$i_r(p) = \sum\limits_{i=1}^{2p-r} d(x_i,x_{i+r})/(2p-r)$$

$$\leq \frac{\sum\limits_{i=1}^{p_0-r} d(x_i,x_{i+r})}{p} + \frac{\sum\limits_{i=p_0-r+1}^{p_0} d(x_i,x_{i+r})}{p} + \frac{\sum\limits_{i=p_0+1}^{2p-r} d(x_i,x_{i+r})}{2p-r}$$

$$\leq \varepsilon'' \quad \frac{\sum\limits_{i=p_0-r+1}^{p_0} d(x_i,\overline{y_{i+r}})}{p} + \frac{\sum\limits_{i=p_0-r+1}^{p_0} d(\overline{y_{i+r}},x_{i+r})}{p}$$

$$+ \frac{\sum\limits_{i=p_0+1}^{2p-r} d(x_i,x_{i+r})}{2p-r}$$

$$\leq \varepsilon'' + \varepsilon' + \frac{p_0-p_0+r-1+1}{p} \; \varepsilon + \frac{\sum\limits_{i=p_0+1}^{2p-r} d(x_i,\overline{y_i}) + d(\overline{y_i},x_{i+r})}{2p-r}$$

(r   is a multiple of k, hence : $\overline{i+r} = \overline{i}$;

hence $y_{\overline{i+r}} = y_i$)

$$\leq \varepsilon'' + \varepsilon' + \varepsilon + 2\varepsilon = \varepsilon_0/16 + \varepsilon_0/16 + \varepsilon_0/8 + \varepsilon_0/4 = \varepsilon_0/2 \qquad (**)$$

The relation (*) and (**) prove that, for $p \geq p'$, (P_k) holds.

If we want to determine the period of a sequence, the lemma suggests that we compute the coefficients $i_r(p)$ (p fixed, $r \in \{1,2,...,p\}$) the smallest gives either the period or multiple of the period. The difficulty is now to distinguish when we have multiples. The various ideas for solving this problem give rise to various algorithms.

Before the precise descriptions, let us explain the principles.

Assume, for example (see table 1, p. 56) that we have points $x_0, x_1, ..., x_{20}$ (p = 10) : We can compute $i_1(10)$, $i_2(10), ..., i_{10}(10)$. We obtain a diagram as in the following :

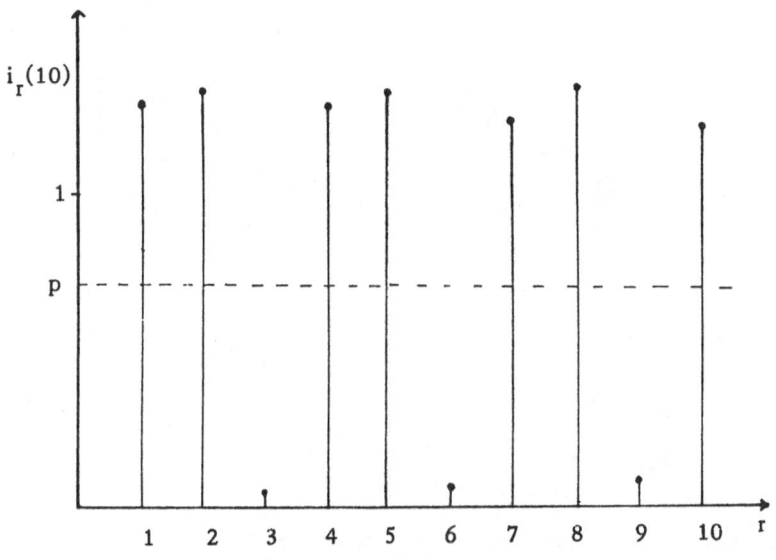

Figure 1

From lemma 1, the corfficients are $i_3(10)$; $i_6(10)$, $i_9(10)$ the smallest have indices which are multiples of the period (in this example 3).

The simplest idea is to propose the index r of the smallest $i_r(10)$, $r \in \{1,2,...,10\}$ for the answer, but lemma 1 tells us that this is only a multiple of the period. If we have additional information, for example that k is a prime number (k $\in$ M = $\{2,3,5,7,11,...\}$) , then this method becomes interesting (see theorem 2). This is the idea of algorithms $I_1(M)$.

Without additional information on k , lemma 1 suggests that we check a line (the dotted line of Figure 1), dividing the points into two sets, the lower one being a set of multiples of a certain integer.

Concerning our particular example, there is only one possible partition, the corresponding integer being k = 3.

This is the idea for algorithm I2, which suggests that we choose the smallest integer k ≥ 2 such that there exists a dotted line as described.

Without information concerning the period k , it may be necessary to know something about the mutual distance between accumulation points of the considered sequence.

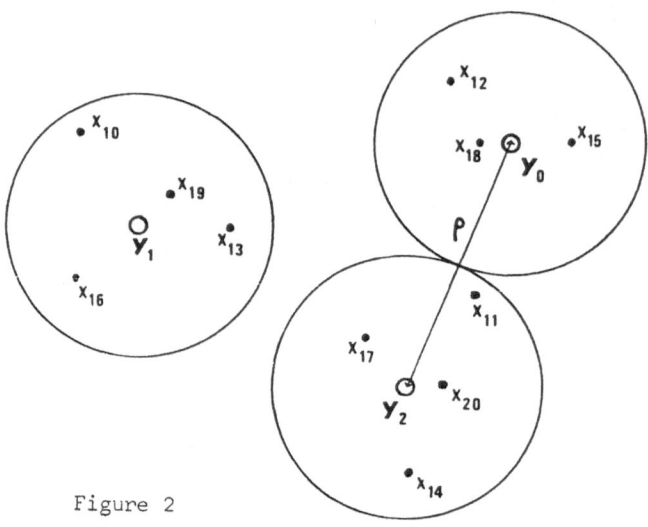

Figure 2

If the mutual distance is greater than $\rho \in R^{+*}$, from lemma 1 we know that after a certain step all the coefficients corresponding to multiples of k are smaller than ρ , and all the others are greater than ρ . If we draw an horizontal line on ρ and if we propose at step p the smallest index r of the $i_r(p)$ staying below this line, we shall obtain the period at the end. This is the idea of algorithm I3(p) .

The algorithm is based on a similar idea, but ρ is now depending on $i_1(p)$ (for example, one may choose $\rho = i_1(p)/10$).

Algorithm I1(M) (M is a fixed subset of N^*).

Step p :

Points $x_1, x_2, \ldots, x_{2p}$ can be used.

$i_r(p)$ is computed for every $r \in M \cap \{1,2,\ldots,p\}$. The proposed answer is N1(p) = j with :

$$i_j(p) = \min\{i_{j'}(p) \mid j' \in M \cap \{1,2,\ldots,p\}\}$$

Algorithm I2

Step p :

Points $x_1, x_2, \ldots, x_{2p}$ can be used.

$i_1(p)$, $i_2(p)$, $\ldots$, $i_p(p)$ are computed. The proposed answer is

$\quad N2(p) = \min\{j \in \{2,3,\ldots,p\} \mid (P_j)\}$

(if there is no j satisfying (P_j) , we set $N2(p) = 1$)

Algorithm I3(ρ) (ρ is a fixed positive parameter)

Step p :

Points $x_1, x_2, \ldots, x_{2p}$ can be used.

$i_1(p)$, $i_2(p)$, $\ldots$, $i_p(p)$ are computed. The proposed answer is

$\quad N3(p) = \min\{j \in \{1,2,\ldots,p\} \mid i_j(p) \leq \rho\}$

(if there is no j such that $i_j(p) \leq \rho$, then we set $N3(p) = 1$).

Algorithm I4(ε) (ε is a fixed positive parameter)

Step p :

Points $x_1, x_2, \ldots, x_{2p}$ can be used.

$i_1(p)$, $i_2(p)$, $\ldots$, $i_p(p)$ are computed. The proposed answer is

$\quad N4(p) = \min\{j \in \{1,2,\ldots,p\} \mid i_j(p) < \varepsilon\, i_1(p)\}$

(if there is no j such that $i_j(p) \leq \varepsilon\, i_1(p)$, then we set $N4(p) = 1$).

Let $M \subset N^*$, $\rho \in R^{+*}$, $\varepsilon \in R^{+*}$ we set :

$\quad \mathrm{APer}^*_M(E) = \bigcup_{m \in M} \mathrm{APer}^*_m(E)$;

$\quad \mathrm{APer}^{**}(E) = \{(x_n) \in \mathrm{APer}^*(E) \mid$ the period of (x_n) is ≥ 2 and

$\qquad\qquad \not\exists\, n_0 \in N ; \not\exists\, !\; j \geq 2 , \;\forall\, n \geq n_0\; (P_j)\}$

$\quad \mathrm{APer}^*_\rho(E) = \{(x_n) \in \mathrm{APer}^*(E) \mid \rho < \min\{d(y_i, y_j) \mid i \neq j\}\}$

$\quad \mathrm{APer}^*_{[\varepsilon]}(E) = \{(x_n) \in \mathrm{APer}^*(E) \mid$

$\qquad\qquad \min\{d(y_i, y_j) \mid i \neq j\} \geq \varepsilon\, \max\{d(y_i, y_j) \mid i \neq j\}\}$

Proposition 3

Q is the question "what is the period of (x_n) ?"

Let $M \subset N^*$, $\rho \in R^{+*}$, $\varepsilon \in R^{+*}$

(i)   I1(M) is satisfactory for Q on $APer^*_M(E)$ if M
does not contain two integers, one of which is a
multiple of the other,

(ii) I2 is satisfactory for Q on $APer^{**}(E)$,

(iii) I3(ρ) is satisfactory for Q on $APer^*_\rho(E)$,

(iv) I4(ε) is satisfactory for Q on $APer^*_{[\varepsilon]}(E)$.

Proof

(i)   Let $(x_n) \in APer^*_M(E)$. Let $k \in M$ be the period of (x_n).
From (ii) of lemma 1, after a certain step p_1 , the property
(P_k) is satisfied, hence we have N1(p) = k.

(ii) Let $(x_n) \in APer^{**}(E)$ and let k be the period of (x_n).
From (ii) of lemma 1, there exists p_1 such that
$\forall p \geq p_1 : (P_k)$.
From the hypothesis, there exists p_2 such that :

$$\forall p \geq p_2 : \exists ! j : (P_j) ,$$

hence for every $p \geq \max\{p_1, p_2\}$ we have N2(p) = k.

(iii) Let $(x_n) \in APer^*_\rho(E)$. Let k be the period of (x_n).
From (i) of lemma 1, we have :

$$\left| \begin{array}{l} \lim_{p \to \infty} i_k(p) = 0 \\ r < k \Rightarrow \lim_{p \to \infty} i_r(p) \geq \varepsilon_0 = \min\{d(y_i, y_j) | i \neq j\} > \rho \end{array} \right.$$

Thus, after a certain step p_1 , we obtain :

$$\left| \begin{array}{l} i_k(p) \leq \rho , \\ r < k \Rightarrow i_r(p) > \rho . \end{array} \right.$$

So that N3(p) = k.

(iv) Let $(x_n) \in APer^*_{[\varepsilon]}(E)$. Let k be the period of (x_n).
From lemma 1 (i), we have :

$$\lim_{p \to \infty} i_k(p) = 0;$$

$$r < k = \lim i_r(p) \geq \varepsilon_0 = \min\{d(y_i, y_j) \mid i \neq j\}$$

$$\leq \varepsilon_1 = \max\{d(y_i, y_j) \mid i \neq j\}.$$

Thus, after a certain step p_1, we obtain :

$$r < k \Rightarrow \begin{vmatrix} i_r(p) - \varepsilon \, i_1(p) \geq \dfrac{\varepsilon_0 - \varepsilon\varepsilon_1}{2} > 0 \\[2mm] \varepsilon \, i_1(p) - i_k(p) \geq \dfrac{\varepsilon\varepsilon_0}{2} > 0 \, ; \end{vmatrix}$$

and so $N4(p) = k$.

Remarks

1) If after a certain step, it seems that $g.c.d.(N1(q) \mid q \geq p)$ is constant, then $M(p)$ is a multiple of the period, thus if $M(p)$ is prime, we are sure that it is the period. In particular, if $N1(p)$ is regularly a fixed prime member p , then this prime number is the period.

2) If the space E is discrete and if there is $\rho \in R^{+*}$ such that $\forall \; x, y \in E^2 : x \neq y \Rightarrow d(x, y) > \rho$ (if E is finite this property holds), then algorithm $I3(\rho)$ is satisfactory for every sequence $(x_n) \in APer^*(E)$ (this does not contradict proposition 5, which does not hold when E is too simple : see remark 1 of proposition 1).

Hence, if we want to determine the period of a sequence $(x_n) \in UPer^*(E)$, we can take the discrete distance on E ($d(x, y) = 0$ if $x = y$, $d(x, y) = 1$ if $x \neq y$) and then apply $I3(\rho)$ with $\rho < 1$.

Numerical experiments have been performed. In particular, we have considered sequences given by $x_{n+1} = f(x_n)$ with the function $f_{k,g}$ defined by :

$$f_{k,g}(x) = g(x - [x+1/2]) + \overline{\frac{[x+1/2]) + 1}{}}^{k}$$

where $[x]$ is the integer part of x and where g is a contractant function from $[-1/4 \, , \, +1/4]$ into itself.

Such $f_{k,g}$ are continuous on $A = \underset{0 \leq i < k}{\cup} [i-1/4 \, , \, i+1/4]$, and for every $x_0 \in A$, the sequence $x_{n+1} = f(x_n)$ is asymptotically periodic with period k.

With $k = 5$, $g(x) = x^{1.1}$, $x_0 = 0$, the first points are :

Table n° 1[1]			
n	x_n	n	x_n
0	0,2	10	1,01538
1	1,17026	11	2,01013
2	2,14264	12	0,00640
3	0,11740	13	1,00386
4	1,09476	14	2,00221
5	2,07486	15	0,00120
6	0,07777	16	1,00061
7	1,04344	17	2,00029
8	2,03174	18	0,00012
9	0,02248	19	1,00005

At the step p , the detector coefficients are :

Table n° 2[1]			
j	$i_j(10)$	j	$i_j(10)$
1	1,31277	6	0,04681
2	1,34114	7	1,28936
3	0,02238	8	1,36180
4	1,30141	9	0,06732
5	1,35180	10	1,27372

(1)   Computation made par M. LE HELLOCO on a P.E.T. Commodore micro-computer.

If we classify the $i_j(10)$, we have :

$i_3(10) \leq i_6(10) \leq i_9(10) \leq i_{10}(10) \leq i_7(10) \leq i_4(10) \leq i_1(10) \leq i_2(10)$
    $\leq i_5(10) \leq i_8(10)$.

With this example, we see that algorithms I1($\mathbf{N}$) , I2 , I3(ρ) (if $\rho < 1,3$) and I4(ε) (if $\varepsilon < 1$) give the correct answer.

However, other numerical experiments allow us to claim :

— I1($\mathbf{N}$) may be wrong, especially with sequences which oscillate around their accumulation points.

— I2 is more often satisfatory but may be wrong, especially if the dimension is ≥ 2.

— It is always posssible to make I3(ρ) good, but the difficulty is to choose ρ.

— It is always possible to made I4(ε) good and $\varepsilon = 0,2$ usually gives good results (sometimes $\varepsilon = 0,1$ is necessary). From the numerical point of view, it seems that I4(ε) is the best of the four.

b) Methods using barycenters

We need the following definition to build our algorithms :

Definition

Let $(x_n, x_{n+1}, \ldots, x_m)$ be a finite sequence of elements of a normed linear space (the distance associated with the norm is denoted by d).

Let $\varepsilon \in \mathbf{R}^+$, $k \in \mathbf{N}$. We say that k is a period within ε of $(x_n, x_{n+1}, \ldots, x_m)$ if and only if :

$$(P_{k,\varepsilon}) \quad \left| \begin{array}{l} \forall\ i \in \{0, 1, \ldots, k-1\} \\[2mm] \forall\ j \in \{n, n+1, \ldots, m\} \end{array} \right. \quad \bar{j}^k = i \Rightarrow d(x_j, G(k,i)) \leq \varepsilon\ ,$$

where :

$$G(k,i) = \sum_{j=n}^{m} x_j / \text{card}\{j \mid n \leq j \leq m,\ \bar{j} = i\}$$

$$\bar{j} = i$$

(picking one point every k points, we obtain k sets, then we compute their barycenters, then we test if the points of a set are near the corresponding barycenter).

Remark

If $k \geq m-n+1$, k is the period within 0 (hence within ε for $\varepsilon \in \mathbf{R}^+$) of the sequence $(x_n, x_{n+1}, \ldots, x_m)$. Thus only periods $< m-n+1$ are of interest.

Example

Consider the sequence (0,6,15,1,4,14,0,5,16)

. 3 is a period within 1   for

$$G(3,0) = \frac{1}{3} \; , \; G(3,1) = 5 \; , \; G(3,2) = 15.$$

. 6 is also a period within 1.

. However, 1,2,4,5,7,8, are not periods within 1.

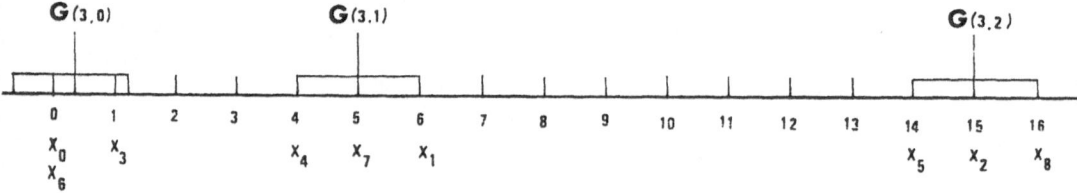

Figure 3

Algorithm J1(ε(p),α(p),β(p))

(ε(p)) is a sequence of real numbers converging to 0 (i.e.
ε(p) $\in$ Conv$_0$($\mathbf{R}^+$));

(α(p)) and (β(p)) are two sequences of integers such that

$$\lim_{p \to \infty} \alpha(p) = +\infty \; , \; \beta(p) \geq \alpha(p)$$

Step p

Points $x_{\alpha(p)}$, $x_{\alpha(p)+1}$,...,$x_{\beta(p)}$ can be used.

We propose the answer N1(p) , the smallest period within ε(p) of
the sequence ($x_{\alpha(p)}$, $x_{\alpha(p)+1}$,...,$x_{\beta(p)}$)

Algorithm J2(ε,α(p),β(p))

ε is a fixed real number;
α(p) and β(p) are two sequences of integers such that :

$$\lim_{p \to \infty} \alpha(p) = +\infty \; , \; \beta(p) \geq \alpha(p)$$

Step p

Points $x_\alpha(p)$, $x_\alpha(p)+1$,...,$x_\beta(p)$ can be used.

We propose the answer N2(p), the smallest period within ε of the sequence $(x_\alpha(p),x_\alpha(p)+1,...,x_\beta(p))$.

Let $(\varepsilon(p)) \in \mathbf{Conv}_0(\mathbf{R}^{+*})$, $\varepsilon \in \mathbf{R}^{+*}$ we set :

  $\mathbf{APer}_{\varepsilon(p)}(E) = \mathbf{APer}(E) \cap \mathbf{Fini}_{\varepsilon(p)}(E)$,

  $\mathbf{APer}_\rho(E) = \mathbf{APer}(E) \cap \mathbf{Fini}_\rho(E)$.

Proposition 4

Let Q be the question "what is the period of the sequence (x_n) ?"

Let $\varepsilon(p) \in \mathbf{Conv}_0(\mathbf{R}^+)$, $\rho \in \mathbf{R}^{+*}$.

(i) $J1(\varepsilon(p),p,2p)$ is satisfactory for Q on $\mathbf{APer}_{\varepsilon(p)}(E)$

(ii) $J2(\rho/5,p,2p)$ is satisfactory for Q on $\mathbf{APer}_\rho(E)$.

Proof

(i) Let $(x_n) \in \mathbf{APer}_{\varepsilon(p)}$ and let k be its period. We denote by $y_0,y_1,...,y_{k-1}$ the limits of

  $(x_{nk}),(x_{nk+1}),...,(x_{nk+k-1})$,

  and we denote by $z_1,z_2,...,z_{k'}$ the accumulation points of (x_n) $(k' \le k)$.

  From the hypothesis, there exists p_1 such that :

  $\forall\ p \ge p_1 : d(x_p,x_{\overline{p}}) \le \varepsilon(p)$.

  If $p \ge p_1$, then :

  $$d(G(k,i),y_i) \le \sum_{j=p}^{2p} d(x_j,y_i)/\mathrm{card}\{j|p \le j \le 2p ,\ \overline{j} = i\} ,$$

  $$\overline{j}=i$$

  Thus $(P_{k,\varepsilon(p)})$ is satisfied. Hence, after step p_1, k is the period within $\varepsilon(p)$ of $(x_p,...,x_{2p})$.

  Now, let $r < k$. We have to prove that after a certain step p, r is not a period within $\varepsilon(p)$ of $(x_p,x_{p+1},...,x_{2p})$.

One of the sequence $(x_{nr}),(x_{nr+1}),\ldots,(x_{nr+r-1})$ is not convergent, say x_{nr+i}. This sequence is asymptotically periodic with period at most k.

Let $z_1,z_2 \in \{y_0,y_1,\ldots,y_{k-1}\}$, $z_1 \neq z_2$, be two of its accumulation points. We set $\varepsilon_1 = d(z_1,z_2)$.

Let $p_1 \in \mathbf{N}$ such that :

$$\left| \begin{array}{l} p_1 \geq k \\ \forall\ p \geq p_1 : \varepsilon(p) \leq \varepsilon_1/4 \ \text{and}\ d(x_p,y_{\bar{p}}) \leq \varepsilon(p). \end{array} \right.$$

If $p \geq p_1$, we have :

$\nexists\ j_1 \in \{p,p+1,\ldots,2p\} : \bar{\jmath}^{r}_1\ :\ i\ \text{and}\ d(x_{j_1},z_1) \leq \varepsilon_1/4$

$\nexists\ j_2 \in \{p,p+1,\ldots,2p\} : \bar{\jmath}^{r}_2\ =\ i\ \text{and}\ d(x_{j_2},z_2) \leq \varepsilon_1/4.$

Thus either j_1 or j_2 satisfies :

$d(x_j,G(r,i)) \geq \varepsilon_1/2 > \varepsilon(p).$

This proves that :

$$\left| \begin{array}{l} \nexists\ i \in \{0,1,\ldots,k-1\} \\ \nexists\ j \in \{p,p+1,\ldots,2p\} \end{array} \right. \qquad \bar{\jmath}^r = i\ \text{and}\ G(r,i) > \varepsilon(p)$$

Thus, at stage p (if $p \geq p_1$) , r is not a period within $\varepsilon(p)$ of $(x_p,x_{p+1},\ldots,x_{2p})$.

(ii) The proof is similar.

Remarks

1) Proposition 4 remains true if, instead of the algorithms $J1(\varepsilon(p),p,2p)$ and $J2(\varepsilon,p,2p)$, we consider the algorithms $J1(\varepsilon(p),\alpha(p),\beta(p))$ and $J2(\varepsilon,\alpha(p),\beta(p))$ with :

$$\lim_{p\to\infty} \alpha(p) = \lim_{p\to\infty} (\beta(p) - \alpha(p)) = \lim_{p\to\infty} \beta(p-1) - \alpha(p)) = +\infty$$

2) It is possible to define another notion of period within ε for a finite sequence, by replacing $(P_{k,\varepsilon})$ by :

$$(P'_{k,\varepsilon}) \qquad \left| \begin{array}{l} \forall\ j_1 \in \{n,n+1,\ldots,m\} \\ \forall\ j_2 \in \{n,n+1,\ldots,m\} \end{array} \right. \qquad \bar{\jmath}_1^k = \bar{\jmath}_2^k => d(x_{j_1},x_{j_2}) \leq \varepsilon$$

With this definition (avoiding calculations of barycenters but involving more tests : a ≤ b), we obtain the same results. if we use this
alternative definition we can also replace the normed linear space   E
by a metric space.

Various numerical experiments have been done using algorithms   J1   and
J2 . We have particularly considered sequences with a number of accumulation points less than the period   (k' < k).

Examples

Consider the following sequence :

$$x_n = \left[(n^{-7}+1)/2 \right] + 1/(n+5)$$

which has period 7. Its subsequences (x_{7n}) , (x_{7n+1}) , ... , (x_{7n+6})
converge respectively to 0,1,1,2,2,3 and 3.

Step   p = 21   of algorithm   J2 (1/5,p,p+19)   gives the calculation :

$$
\begin{array}{ll}
G(2,1) = 1,9392 & G(7,0) = 0,0319 \\
G(3,0) = 1,9055 & G(7,1) = 1,0309 \\
G(4,1) = 1,8649 & G(7,2) = 1,0299 \\
G(5,1) = 1,8322 & G(7,3) = 2,0291 \\
G(6,3) = 2,6314 & G(7,4) = 2,0282 \\
G(7,6) = 3,0330 & G(7,5) = 3,0229
\end{array}
$$

The calculations of   G(2,0) , G(3,1) , G(3,2) , G(4,0) , G(4,2) ,
G(4,3) , G(5,0) , G(5,2) , G(5,3) , G(5,4) , G(6,0) , G(6,1) , G(6,2) ,
G(6,4) , G(6,5)   are not necessary.

The smallest period within   ε = 1/5   is   7.   It is the correct answer.

Even when   k'   is small relative to   k ,   it seems that the calculation of the smallest period of a finite sequence allows us to obtain
the period   k .

For example, for $x_n = \left[n^{20}/20 \right] + 1/(n+5)$ having subsequences
$(x_{20n}),(x_{20n+1}),...,(x_{20n+19})$ converging to   (0,0,...,0,1,) ,   the
calculation of the smallest period within   1/5   of $(x_{20},...,x_{200})$
gives   20   which is the correct answer.

Similarly, for the sequence :

$$x_n = E(n^{27}/27) + 1/(n+5) + n^{-3} ,$$

having subsequences $(x_{27n}),(x_{27n+1}),...,(x_{27n+26})$ converging to
(0,1,2,0,1,2,0,1,2,....,0,1,2,0,1,3)   the smallest period within   1/5
of $(x_{20},...,x_{200})$ is   27   which is the correct answer.

These results show that practically, the notion of smallest period
within   ε   (and hence algorithms J1, J2) is very efficient.

c) Limitation results

Proposition 5

Let $M \subset N^*$. If M contains two integers, one a multiple of the other, then the question "what is the period of (x_n) ?" is undecidable in the limit on $APer^*_M(R^m)$.

$APer^*_M(R^m)$ is the family of sequences defined by :

$$(x_n) \in APer^*_M(R^m) \iff \left| \begin{array}{l} (x_n) \in APer^*(R^m) \text{ and} \\ \text{the period of } (x_n) \text{ is} \\ \text{an integer of } M . \end{array} \right.$$

Proof

Let $k \in N^*$, $r \in N^*$; $r \neq 1$ such that $k \in N$ and $kr \in N$.

For convenience of notation, we use $m = 2$. Assume there exists a normal algorithm satisfactory in the limit on $APer^*_M(R^2)$ for the question of the period.

Let (x^0_n) be the following sequence :

$$(x^0_0, x^0_1, \ldots, x^0_n, \ldots) = ((1,0),(2,0),\ldots,(r,0),\ldots)$$

The last three dots meaning that we periodically take the same r points.

From the hypothesis, there is n_0 such that $N^0(n_0) = r$ and $r|(n_0+1)$ (we denote by $N^0(p)$ the sequence of answers given by A for (x^0_n)).

Let (x^1_n) be the sequence defined by :

$$x^1_n = x^0_n \text{ if } n \leq n_0$$

$$(x^1_{n_0+1}, x^1_{n_0+2}, \ldots) =$$

$$((1,\overset{1}{\underset{2}{-}}),(2,\overset{1}{\underset{2}{-}}),\ldots,(r,\overset{1}{\underset{2}{-}}),,(1,\overset{2}{\underset{2}{-}}),(2,\overset{2}{\underset{2}{-}}),\ldots,(r,\overset{2}{\underset{2}{-}}),(1,\overset{3}{\underset{2}{-}}),\ldots,$$

$$(1,\overset{k}{\underset{2}{-}}),\ldots,(r,\overset{k}{\underset{2}{-}}),\ldots) ,$$

the last three dots meaning that we periodically take the same rk points).

This sequence is asymptotically periodic of period rk , thus there exists n_1 , such that $N^1(n_1) = rk$ and $r|(n_1+1)$ (we denote by $N^1(p)$ the sequence of answers given by A for (x^1_n)).

Let $(x^2{}_n)$ be the sequence defined by :

$x^2{}_n = x^1{}_n$ if $n \leq n_1$

$(x^2{}_{n_1+1}, x^2{}_{n_1+2}, \ldots) = ((1,0),(2,0),\ldots,(r,0),\ldots)$,

the last three dots meaning that we periodically take the same r points.

We continue in the same way :

$(x^{i+1}{}_{n_i+1}, x^{i+1}{}_{n_i+2}, \ldots) = ((1,0),(2,0),\ldots,(r,0),\ldots)$ if i is odd

$(x^{i+1}{}_{n_1+1}, x^{i+1}{}_{n+2}, \ldots) =$

$$= ((1, \frac{1}{2^{i+1}}),(2, \frac{1}{2^{i+1}}),\ldots,(r,\frac{1}{2^{i+1}}),(1,\frac{2}{2^{i+1}}),$$

$$(2,\frac{2}{2^{i+1}}),\ldots,(r,\frac{2}{2^{i+1}}),(1,\frac{3}{2^{i+1}}),\ldots,(1,\frac{k}{2^{i+1}}),\ldots,(r,\frac{k}{2^{i+1}}),\ldots)$$

if i is even.

Then we consider the sequence :

$$(x_n) = (x^0{}_0, x^1{}_0, \ldots, x^0{}_{n_0}, x^1{}_{n_0+1}, \ldots, x^1{}_{n_1}, x^2{}_{n_1+1}, \ldots)$$

This sequence is asymptotically periodic of period r ; however :

$N(n_1) = rk$, $N(n_3) = rk, \ldots$.

Remarks

1) Proposition 5 is still true if we replace R by any metric space such that $E^{\alpha\alpha} \neq \emptyset$ (E^α denotes ther set of accumulation points of E and $E^{\alpha\alpha} = (E^\alpha)^\alpha$).

2) The result of proposition 5 is truly precise : in particular, it says :

* it is impossible ot distinguish in the limit between a sequence which is asymptotically convergent of period 2 and a sequence which converges.

* it is not sufficient to know that a sequence has only a finite number of accumulation points, each with a strictly positive strength, in order to have the existence of algorithms determining the period.

3) Using proposition 2 (i) , we obtain that the condition "M does not contain two integers one a multiple of the other" is necessary and sufficient for the existence of algorithms determining the period of sequences of $APer^*_M(R^m)$.

The next theorem is a summation of the results of propositions 3,4 and 5.

Theorem 6

The question "What is the period of the sequence (x_n) ?" is decidable in the limmit on :

(i) $APer^*_M(E)$ for every $M \subset N^*$ which does not contain two integers, one a multiple of the other;

(ii) $APer^{**}(E)$;

(iii) $APer_\rho(E)$ for every $\rho \in R^*$;

(iv) $APer^*_{[\varepsilon]}(E)$ for every $\varepsilon \in R^{+*}$;

(v) $APer_{\varepsilon(p)}$ for every $\varepsilon(p) \in Conv_0(R^+)$;

and is undecidable in the limit on :

(j) $APer^*_M(R^m)$, if $M \subset N^*$ contains two integers, one a multiple of the other.

5 - FAMILIES OF SEQUENCES OF ITERATIONS

Now we consider the set IE^N of sequence (x_n) , $x_n \in E$ where (x_n) is generated by a continuous function $f : E \to E$:

$$x_n \in IE^N \iff \begin{cases} \text{there exists a continuous application } f : E \to E \\ \\ \text{such that } \forall n \in N \quad x_{n+1} = f(x_n) \end{cases}$$

We restrict ourselves to the case $E = [0,1]$, and give some generalizations.

We use the following notation :

$IStati(E) = (IE^N) \cap Stati(E)$;
$IPer(E) = (IE^N) \cap Per(E)$;
$IUPer(E) = (IE^N) \cap UPer(E)$;
etc...

Three characterisations have to be noted :

(a) (x_n) is ultimately $\quad\Bigg|\quad$ <=> $(\exists\ m \in M\ ,\ x_n = x_{m+1})$

      stationary

(b) (x_n) is ultimately $\quad\Bigg|\quad$ <=> $(\exists\ m,p \in N\ ,\ m \neq p$ and

      periodic $\qquad\qquad\qquad\qquad\qquad x_m = x_p)$

(c) (x_n) is asymptotically $\quad\Bigg|\quad$

      periodic with period p $\qquad$ <=> $((x_n)$ has p   accumulation
                                 points)

We consider anew the theorems of sections 2,3 and 4. Because of the
additional information that (x_n) is a sequence of iterations, the
positive part of the theorems increases. Because of (c) (which means
$\text{IFini}_p(E) = \text{IAPer}_p(E)$)   certain parts of the theorems lose their
meanings;. When a negative result remains true it needs a new proof,
for we have to show that every sequence used is a sequence of iterations (for example, see the proof of the lemma 2).

Theorem 7

The question "is (x_n) a convergent sequence ?" is decidable in the
limit.

  (i)   on **IUPer** $\left[0,1\right]$ ∪ **IConv**$\left[0,1\right]$;

  (ii) on **IAPer**$_\varepsilon\left[0,1\right]$ for every $\varepsilon > 0$;

  (iii) on **IUStati**$\left[0,1\right]$ ∪ **Turb**$\left[0,1\right]$;

      and is undecidable in the limit :

  (j)   on **IAPer**$\left[0,1\right]$;

  (jj) on **IConv**$\left[0,1\right]$ ∪ **Turb**$\left[0,1\right]$.

Remarks

1) The theorem is still true for every dense subset of $\left[0,1\right]$, in
particular for **Q** ∩ $\left[0,1\right]$, for $\left[0,1\right]$ ∩ R_c (R_c : the field of calculable reals).

2)   If instead of sequences generated by continuous functions, we
consider only sequences generated by Lipschitz functions, it is likely
that the positive part of theorem 7 increases, indeed we notice that
all the functions used in the proof of (j) and (jj) are
non-Lipschitz.

Proof

(i) We define a normal algorithm $N = (f_n)$ by :

$$f_i(x_0,x_1,\ldots,x_i) = NO \Longleftrightarrow \left| \begin{array}{l} \exists\ j,k \in \{0,1,\ldots,i\} : x_j = x_k \text{ and} \\[2mm] \exists\ n,m \in \{[i/2],[i/2] + 1,\ldots,i\} : \\ x_n \neq x_m \end{array} \right.$$

Indeed : * either (x_n) is not ultimately periodic (hence (x_n) is convergent) ant thus from (b) it is impossible to obtain j and k , and thus the given answer is always correct.

* or (x_n) is ultimately periodic, then beyond a certain stage j and k exist. Now n and m exist for i sufficiently large if and only if (x_n) is not ultimately stationary.

(ii) Since $IAPer_\epsilon[0,1] \subset APer_\epsilon(E)$, this is a consequence of (ii) of theorem 2.

(iii) This follows the more general result :

Lemma 1

The question "is (x_n) a convergent sequence ?" is decidable in the limit on $IUPer[0,1] \cup ([0,1]^{\mathbb{N}} - IConv[0,1])$.

Proof

We define a normal algorithm $N = (f_n)$ by :

$f_i(x_0,x_1,\ldots,x_i) = YES \Longleftrightarrow (\exists\ n \in \{0,1,\ldots,i-1\}\ x_n = x_{n+1})$

(j) This follows the more general result :

Lemma 2

The question "is (x_n) a convergent sequence ?" is undecidable in the limit on $IConv[0,1] \cup IAPer_2[0,1]$.

In order to prove this lemma, we need : (α) the notion of insertion of a function on another funtion (β) two particular functions.

(α) Insertion of a function on another function

Let   f,g   be two continuous functions from $[0,1]$ into $[0,1]$; let a,b   satisfy $o \leq a < b \leq 1$.

We define a function h (see figure 4) :

$h(x) = f(x)$ if $x \in [0,1] - [2a-b,2b-a]$;

$h(x) = a + (b-a) \; g(\dfrac{1}{b-a} \, (x-a))$ if $x \in [a,b]$;

$h(x)$ is linear on $[2a-b,a] \cap [0,1]$ and on $[b,2b-a] \cap [0,1]$

$h(x)$ is continuous on $[0,1]$.

We say that h is obtained by the insertion of g on f with respect to the interval [a,b].

Methods similar to insertion have already been used on the study of
iterations ([4],[16],[17] and [32]) .

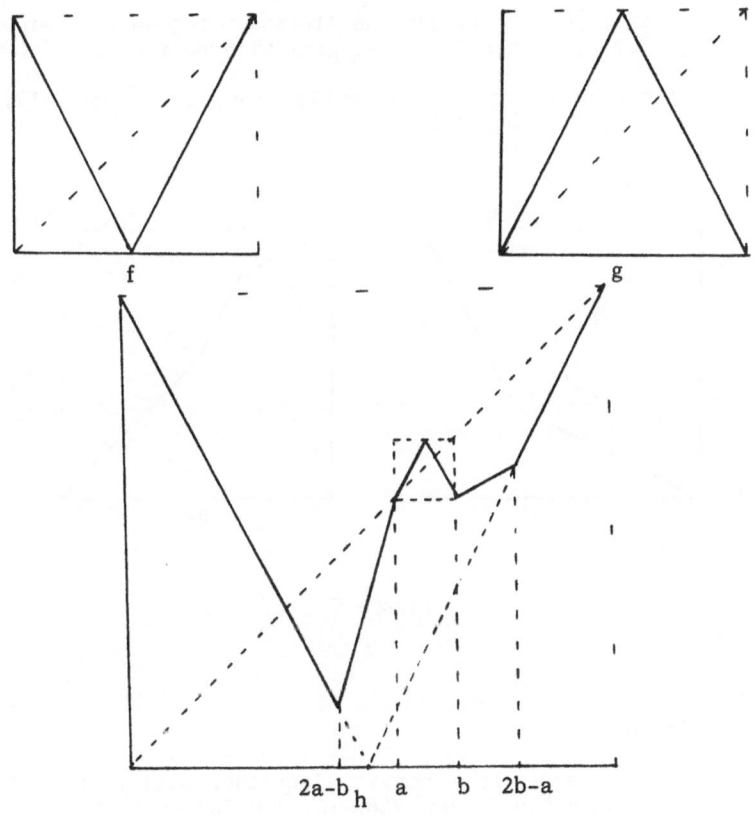

Figure 4

(β) Two particular functions

Let g_0 and g_1 be defined by :

$g_0(x) = x/2$ if $x \in [0,2/3]$; $g_0(x) = 2x-1$ if $x \in [2/3,1]$;

$g_1(x) = 1-x/2$ if $x \in [0,1/3]$; $g_1(x) = 3/2-2x$ if $x \in [1/3,1/2]$;

$g_1(x) = 1-x$ if $x \in [1/2,1]$.

It is easy to verify that :

* for every $x_0 \in [0,1]$, the iteration sequence obtained from x_0 with g_0 is convergent.

* for every $x_0 \in [0,1] - \{1/2\}$ the iteration sequence obtained from x_0 with g_1 is asymptotically convergent with period 2.

(similar functions are frequently used, especially in $[4],[7],[16],[18],[32]$ and $[33]$).

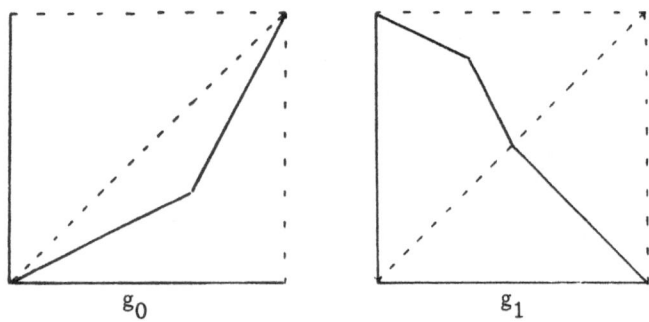

$$g_0 \qquad\qquad\qquad g_1$$

Figure 5

Proof of lemma 2

Assume that $A = (a_i)$ is a normal algorithm, satisfactory for the determination of convergence on **IConv** $[0,1]$ ∩ **IAPer**$_2[0,1]$.

We set $h_0(x) = x$.

Let f_0 be the function obtained by the insertion of g_1 on h_0 with respect ot the interval $[0,1/3]$ (see figure 6).

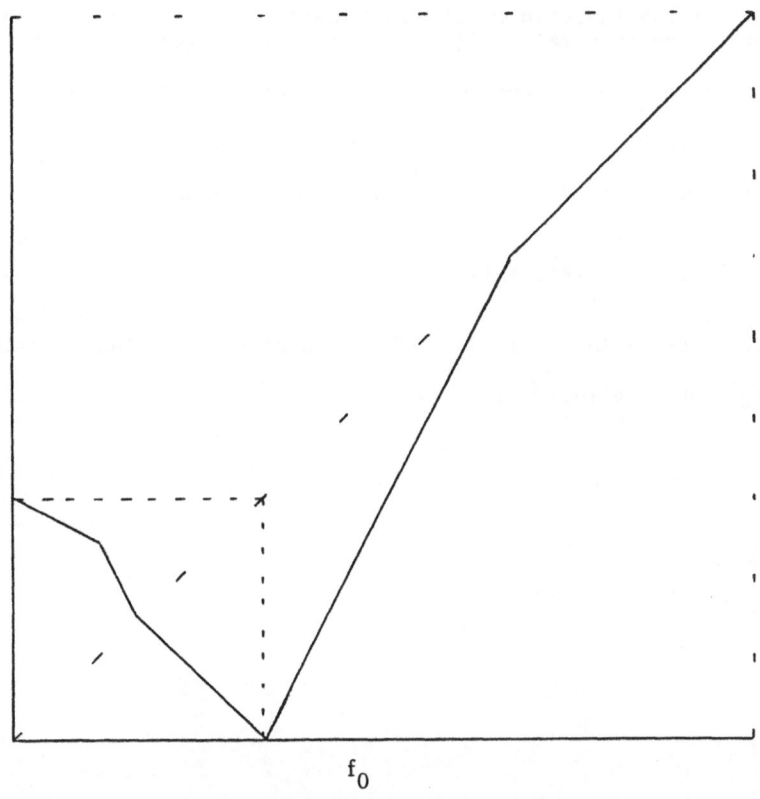

$$f_0$$

Figure 6

Let (x^0_n) be the sequence of iterations of $x_0 = 2/9$ with f_0.

This sequence is asymptotically periodic of period 2; hence there exists $n_0 \in \mathbf{N}$ such that :

$$a_{n_0}(x^0_0, x^0_1, \ldots, x^0_{n_0}) = NO$$

Consider a small interval $]a_1, b_1[$ containing $x^0_{n_0}$ but containing no x^0_i for $i \in \{0, 1, \ldots n_0-1\}$.

We define the function h_1 such that :

$h_1(x^0_{n_0}) = 2/3 + 2/27$:

$h_1(x) = f_0(x)$ on $[0,1] -]a_1, b_1[$;

h_1 is linear and continuous on $[a_1, x^0_{n_0}]$ and on $[x^0_{n_0}, b_1]$.

Let f_1 be the function obtained by the insertion of g_0 on h_1 with respect to the interval $[2/3, 2/3 + 1/9]$ (see figure 7).

Let (x^1_n) be the sequence of iterations of $x_0 = 2/9$ with f_1.

By construction : $x^1_n = x^0_n$ for every $n \in \{0,1,\ldots,n_0\}$.

Again by construction, (x^1_n) is convergent. Hence, there exists n_1 such that :

$$a_{n_1}(x^1_0, x^1_1, \ldots, x^1_{n_1}) = \text{YES}$$

Consider a small interval $]a_2, b_2[$ containing $x^1_{n_1}$ but containing no x^1_i for $i \in \{0,1,\ldots,n_1-1\}$.

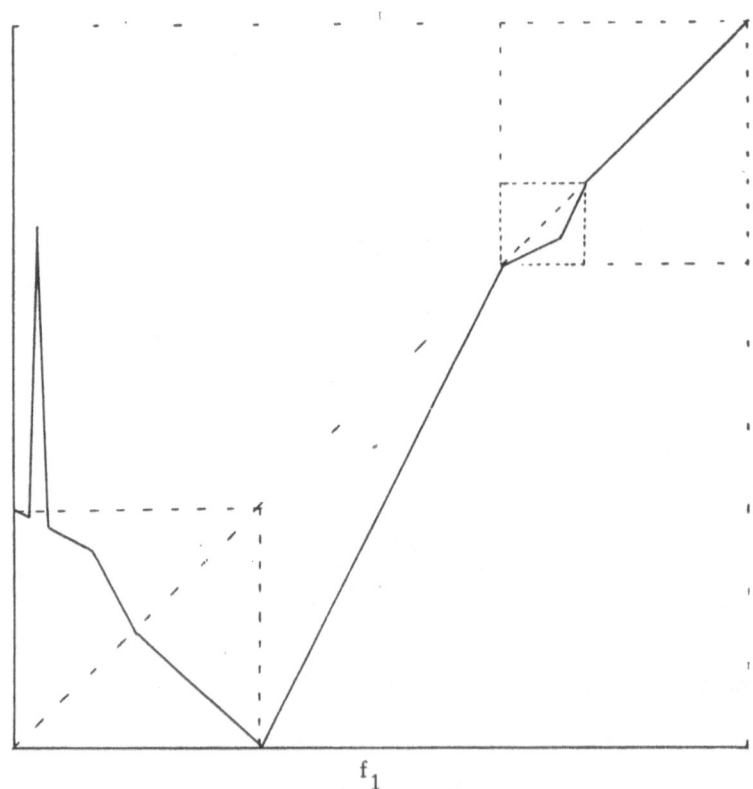

f_1

Figure 7

We define the function h_2 such that :

$$h_2(x^1_n) = 8/9 + 2/81 ;$$

$$h_2(x)^1 = f_1(x) \quad \text{on} \quad [0,1] -]a_2,b_2[;$$

h_2 is linear and continuous on $[a_2, x^1_n]$ and $[x^1_n, b_2]$.

Let f_2 be the function obtained by the insertion of g_1 on h_2 with respect to the interval $[8/9, 8/9 + 1/27]$ (see figure 8).

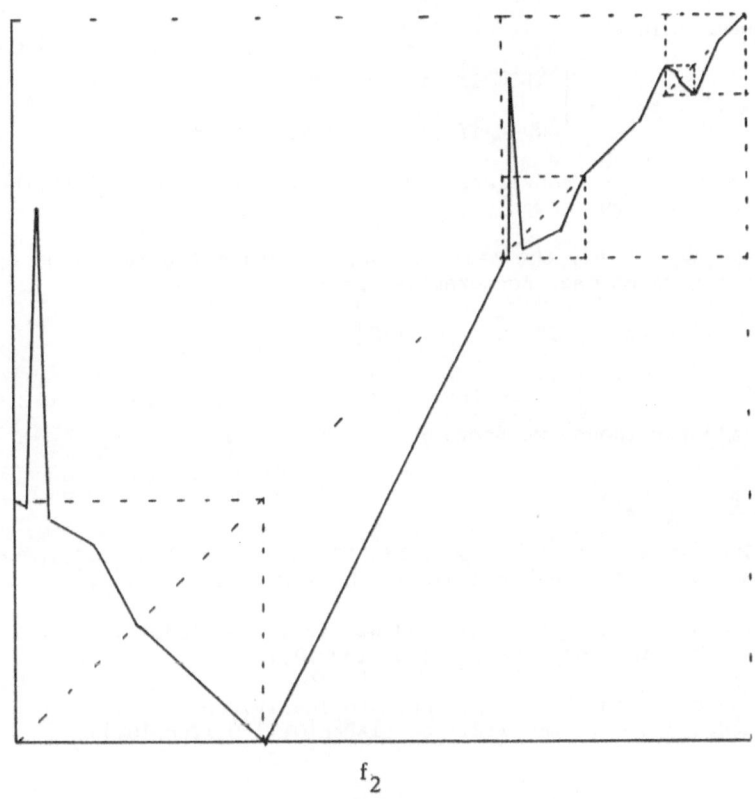

$$f_2$$

Figure 8

Let (x^2_n) be the sequence if iterations of $x_0 = 2/9$ with f_2.

By construction, $x^2_n = x^1_n$ for every $n \in \{0,1,\ldots,n_1\}$.

Again by construction, (x^2_n) is asymptotically periodic of period 2, so there exists n_2 such that :

$$a_{n_2}(x^2_0, x^2_1, \ldots, x^2_{n_2}) = NO$$

Etc...

The sequence of functions (f_i) converges uniformaly to a function f which (by construction) yields the following sequence with $x_0 = 2/9$:

$$(x_n) = (x^0_0, x^0_1, \ldots, x^0_{n_0}, x^1_{n_0+1}, \ldots, x^1_{n_1}, x^2_{n_1+1}, \ldots)$$

Hence, we obtain :

$$a_{n_i}(x_0, x_1, \ldots, x_{n_i}) = \left| \begin{array}{l} YES \quad if \quad i \quad is \ odd \\[2mm] NO \quad\;\; if \quad i \quad is \ even \end{array} \right.$$

The sequence (x_n) converges (to $\ell = 1$) and contradicts the assumptions made about A.

(jj) We use exactly the same method, replacing the function g_1 by a turbulent one. For example g_2 :

$g_2(x) = 2x$ if $x \in [0, 1/12]$

$g_2(x) = 2(1-x)$ if $x \in [1/2, 1]$

Using similar methods, we prove :

Theorem 8

(i) Questions "is (x_n) periodic ?" and "is (x_n) ultimately periodic ?" are decidable in the limit on $I[0,1]^{\mathbf{N}}$.

(ii) The question "is (x_n) asymptotically periodic ?" is decidable in the limit on $\mathbf{IPer}[0,1] \cup \mathbf{ITurb}[0,1]$

(j) The question "is (x_n) asymptotically periodic ?" is undecidable in the limit on $\mathbf{IAPer}[0,1] \cup \mathbf{ITurb}[0,1]$.

Theorem 9

The question "is (x_n) turbulent ?" is decidable in the limit on

(i) $\mathbf{IUPer}[0,1] \cup \mathbf{ITurb}[0,1]$:

and is undecidable in the limit on

(j) $\mathbf{IAPer}[0,1] \cup \mathbf{ITurb}[0,1]$.

Remark

The problem of the decidability in the limit for this question on

$$IAPer_p[0,1] \cup ITurb[0,1] \quad (p \geq 2 \text{ fixed})$$

is not solved. Theorems such as   Th. II-3   of $[8]$ show that a positive answer is perhaps possible.

Theorem 10

The question   "what is the period of (x_n) ?"   is decidable in the limit on :

(i) $IUPer[0,1]$;

(ii) $IAPer_\varepsilon[0,1]$ for every $\varepsilon > 0$;

and is undecidable in the limit on :

(j) $IAPer[0,1]$.

6 - TWO GENERAL RESULTS CONCERNING THE DECIDABILITY IN THE LIMIT

Here we give two general results on problems of decidability in the limit as defined in section 1. The first one claims a sufficient condition for a question to be decidable in the limit on a family   **S** ;   the second one claims a sufficient condition for a question to be undecidable in the limit on a family   **S**.   Unfortunately, none of these conditions are necessary.

Let   Q   be a question having sense on $S \subset E^N$,   and whose set of possible answers is   R .   This question defines a function F_Q from E^N into   R   with   S   as domain of definition, and for which the value on (x_n) is the correct answer of   Q   for (x_n).

We set :

$$S' = \{s \in E^{(N)} | \; \exists \; (x_n) \in S , \; \exists \; n_0 \in N : s = (x_0, x_1, .., x_{n_0})\}$$

$$S'' = S \cup S'.$$

Consider on $E^N \cup E^{(N)}$ the topology defined by the basis of neighborhoods :

* $(x_n) \in E^N$, $x_0 \in N$,

$$U((x_n), n_0) = \{(y_n) \in E^N \mid \forall \; n \leq n_0 : y_n = x_n\}$$

$$\cup \{(y_n) \in E^{(N)} \mid \text{length} (y_n) \geq n_0+1 \quad \text{and}$$

$$\forall \; n \leq n_0 : y_n = x_n\}$$

* $(x'_n) \in E^{(\mathbf{N})}$, $n'_0 \in \mathbf{N}$, $n'_0 + 1 \leq$ length (x'_n)

$U((x'_n),n'_0) = \{(y_n) \in E^{\mathbf{N}} \mid \forall n \leq n'_0 : y'_n = x'_n\}$

$\cup \{(y'_n) \in E^{(\mathbf{N})}$ length $(y'_n) \geq n'_0+1$ and

$\forall n \leq n'_0 : y'_n = x'_n\}$

This topology T_0 is less fine than the topology T_1 defined by completing $E^{(\mathbf{N})}$ with the following distance :

$$d((x_n),(x'_n)) = 1/(m_0+1) ,$$

where m_0 is the largest integer such that :

$m_0+1 \leq$ length (x_n) ,

$m_0+1 \leq$ length (x'_n) ,

$x_n = x'_n$ for every $n \leq n_0$.

Proposition A

(Sufficient condition of decidability in the limit)

If F_Q can be continuously (with respect to T_0 or T_1) extended from S to S'' , then Q is decidable in the limit on S.

Proof

Let F'_Q be the continuous extension of F_Q to S. We define the normal algorithm (f_i) by setting

$$f_i(x_0,x_1,\ldots,x_i) = F'_Q((x_0,x_1,\ldots,x_i))$$

for every $(x_n) \in S$.

Remark

This sufficient condition of decidability in the limit is not necessary. Indeed, let Q be the question "is there a 1 in (x_n)?" defined on $\{0,1\}^{\mathbf{N}}$. Q is obviously decidable in the limit on $\{0,1\}^{\mathbf{N}}$ but F_Q cannot be extended continuously to $\{0,1\}^{\mathbf{N}} \cup \{0,1\}^{(\mathbf{N})}$, for F_Q is not continuous on $\{0,1\}^{\mathbf{N}}$ (consider $(1,0,0,\ldots),(0,1,0,0,\ldots),(0,0,1,0,\ldots)\ldots)$.

Proposition B

(Sufficient condition of non-decidability in the limit).

If the family S satisfies the following condition :

(a) 0) there exists $(x^0_n) \in S$;

    1) for every $m_0 \in N$, there exists $(x^1_n) \in S$ such that :

$$F_Q(x^1_n) \neq F_Q(x^0_n) \quad \text{and}$$

$$\forall \, n \leq m_0 : \quad x^1_n = x^0_n \;;$$

    2) for every $m_1 \geq m_0$, there exists $(x^2_n) \in S$ such that :

$$F_Q(x^2_n) \neq F_Q(x^1_n) \quad \text{and}$$

$$\forall \, n \leq m_1 : \quad x^2_n = x^1_n$$

$$\ldots \quad \ldots \quad \ldots$$

(b) $(x^0_0, x^0_1, \ldots, x^0_{m_0}, x^1_{m_0}+1, \ldots, x^1_{m_1}, x^2_{m_1}+1, \ldots) \in S$

    Then the question Q is undecidable in the limit on S.

Proof

The proof is similar to the proof of lemma 2 of § 2.

Remark

The condition of remanence studied in chapter 5 (see also [22]) which is a sufficient condition of non-accelerability is built on an analogous frame.

APPENDIX 1

Strength of an accumulation point and quickness of a sequence

(a) Strength of an accumulation point

The notion of "strength of an accumulation point" attempts to count the average number of times that the sequence (x_n) returns to a neighborhood of a fixed accumulation point (or for any point of E).

Definition

Let (x_n) be a sequence in the metric space (E,d).

Let $y \in E$. The strength of y with respect to (x_n) is :

$$\alpha(y,(x_n)) = \lim_{\substack{\varepsilon \to 0 \\ \varepsilon > 0}} \liminf_{m \to \infty} \operatorname{card}\{n \in \{0,1,\ldots,m-1\} \mid d(y,x_n) \leq \varepsilon\}/m.$$

(i) $0 \leq \alpha(y,(x_n)) \leq 1$

Indeed, we set :

$$N^{\varepsilon}_m = \operatorname{card}\{n \in \{0,1,\ldots,m-1\} \mid d(y,x_n) \leq \varepsilon\}/m$$

$$N^{\varepsilon} = \liminf_{n \to \infty} N^{\varepsilon}_m$$

We have $0 \leq N^{\varepsilon}_m \leq 1$, hence $0 \leq N^{\varepsilon} \leq 1$ and so

$0 \leq \alpha(y,(x_n)) \leq 1$.

(ii) If (x_n) converges to y then :

$\alpha(y,(x_n)) = 1$ and $\forall z \neq y : \alpha(z,(x_n)) = 0$.

Indeed, let $\varepsilon > 0$, there exists n_0 such that if $m \geq n_0$ then $d(x_m,y) \leq \varepsilon$; hence for every $m \geq n_0$ we have

$(m-n_0)/m \leq N^{\varepsilon}_m \leq 1$. Thus $N^{\varepsilon} = 1$ and $\alpha(y,(x_n)) = 1$.

The other part is similar.

(iii) If $\alpha(y,(x_n)) > 0$, then y is an accumulation point of (x_n) and :

$\forall \varepsilon \in \mathbf{R}^{+*}$, $\exists p_0 \in \mathbf{N}$, $\forall p \geq p_0$

$\{x_p,x_{p+1},\ldots,x_{p^2}\} \cap B(y,\varepsilon) \neq \emptyset$.

We assume that there exists $\varepsilon \in \mathbf{R}^{+*}$ such that

$\forall \, p_0 \in \mathbf{N} \, , \, \exists \, p \geq p_0 \, : \, \{x_p, x_{p+1}, \ldots, x_{p^2}\} \cap B(y, \varepsilon) = \emptyset$

and we try to obtain a contradiction.

We can build a strictly increasing sequence of integers $\{p_n\}$ such that :

$$\{x_{p_n}, x_{p_n+1}, \ldots, x_{p_n^2}\} \cap B(y, \varepsilon) = \emptyset \, ,$$

For every n , we have :

$$N^\varepsilon_{p_n^2} \leq \frac{p_n}{p^2_n} \, ,$$

hence $N^\varepsilon = 0$ and so $\alpha(y, (x_n)) = 0$.

(iv) If $(x_n) \in \mathbf{APer}^*_p(E)$, then every accumulation point y satisfies :
$$\alpha(y, (x_n)) = \frac{1}{p}$$

(v) If $(y_i)_{i \in I}$ is the family of all the accumulation points of (x_n) , then :

$$0 \leq \sum_{i \in I} \alpha(y_i, (x_n)) \leq 1 .$$

Remarks

1) It is not true that if y is an accumulation point of (x_n) then $\alpha(y, (x_n)) > 0$.

For example, with $(x_n) = (0,1,0,0,1,0,0,0,1,\ldots)$ we obtain that :
$\alpha(0, (x_n)) = 1$, $\alpha(1, (x_n)) = 0$.

2) In (v) it is possible to have $\sum_{i \in I} \alpha(y_n, (x_n)) < 1$. For example, with :

$(x_n) = (0,1,1,0,0,0,0,1,1,1,1,1,1,1,1,0,0,\ldots)$

we obtain $\alpha(0, (x_n)) = \alpha(1, (x_n)) = 1/3$.

In (v), it is possible to have $o = \sum$. For example, with

$(x_n) = (0,0,1,0,1,2,0,1,2,3,0,1,2,3,4,\ldots)$,

we obtain $\sum_{m \in \mathbf{N}} \alpha(m, (x_n)) = 0$.

3) A generalization of the notion of strength of an accumulation point was defined and studied in [21].

(b)   Quickness of a sequence

Definition

Let (x_n) be a sequence of points in the metric space E. We denote by $A(x_n)$ its set of accumulation point.

Let $(\varepsilon_n) \in \mathrm{Conv}_0(R^+)$.

We say that (ε_n) is a quickness of the sequence (x_n) if :

$$\exists\, m_0 \in N\ ,\ \forall\, m \geq m_0 : d(x_m, A(x_n)) \leq \varepsilon_m.$$

Example

$$x_n = 1 + 1/(n+1) \quad \text{if} \quad n \ \text{is} \ \text{even},$$

$$x_n = 1/2^n \quad\quad \text{if} \quad n \ \text{is} \ \text{odd},$$

$$(\varepsilon_n) = (1/(n+1)) \quad \text{is a quickness of} \quad (x_n).$$

Remarks

1) It is possible that there is no quickness. For example :

$$(x_n) = (n + (-1)^n \ n)$$

2)   There is no relationship between the notion of strength and the notion of quickness. We can easily find sequences whose accumulation points have a strictly popsitive strength and which are without quickness. Conversely, there exist sequences with quickness and whose accumulation points have a null strength.

Proposition

Let (x_n) be a sequence of points in the metric space E.

 (i)   If

   (T) $\forall\, v \in V(A(x_n))$, $\exists\, n_0 \in N$, $\forall\, n \geq n_0 : x_n \in v$,

   $(V(A(x_n)))$ is the set of neighborhoods of $A(x_n)$),

   then the sequence $\varepsilon_n = d(x_m, A(x_n))$ converges to 0 and is a quickness of the sequence (x_n).

(ii)   If (x_n) has a quickness and if $A(x_n)$ is compact, then (T) is satisfied.

Proof

(i) Let $\varepsilon \in R^{+*}$, set $v = \{x \in E \mid d(x, A(x_n)) < \varepsilon\}$.
     Then v is a neighborhood of $A(x_n)$, hence from (T), there
     exists $m_0 \in N$ such that $m > m_0 \Rightarrow x_m \in v$.
     Hence $d(x_m, A(x_n)) \xrightarrow[m \to \infty]{} 0$.

(ii) Let v be an open neighborhood of $A(x_n)$; cv is closed.
     (cv is the complementary of v)
     Hence : $d(cv, A(x_n)) > 0$
     (here we use the fact that $A(x_n)$ is compact).

     Let n_0 be such that :

$$n < n_0 \Rightarrow \varepsilon(n) < d(cv, (x_n)).$$

     For every $n \geq n_0$, we obtain :

$$x_n \in v .$$

Remark

In (ii) the hypothesis "$A(x_n)$ is compact" can not be relaxed.

Example : $(x_n) = (0,0,1,0,1,2,0,1,2,3,\ldots)$;

$$(y_n) = (x_n + \frac{1}{n+1}) ,$$
$$A(y_n) = N ,$$
$$v = \cup \ ]n - \frac{1}{2^n} , \ n + \frac{1}{2^n}[\ .$$

APPENDIX 2

Decidability in the limit and recursivity

In remark 2 of § 2, we explained, in our definition of decidability in
the limit, why we had not introduced conditions of calculability.

For some examples, we show the way to introduce such conditions and
translate lemmas 1 and 2 of § 2 into this new situation.

In order to speak of calculability, we have to assume that E is
denumerable. We consider only $E = [0,1] \cap Q$ (denoted by $[0,1]_Q$). R
must also be denumerable : we shall assume $R \subset N$.

Let a be a recursive bijection from N into $[0,1]_Q$
(i.e. $a = (a_1, a_2, a_3)$ is a function from N into N^3 such that

$$p \rightarrow (-1)^{a_1(p)} \; a_2(p)/(a_3(p)+1)$$ is a bijection from $\mathbf{N}$ into $[0,1]_\mathbf{Q}$
Let α_i be a recursive bijection from $\mathbf{N}^i$ into $\mathbf{N}$. For example, α_i
may be defined by :

$$\alpha_2(x_1,x_2) = (x^2_1 + 2\, x_1 x_2 + x^2_2 + 3\, x_1 + x_2)/2$$

$$\alpha_{i+1}(x_1,x_2,\ldots,x_{i+1}) = \alpha_2(\alpha_i(x_1,x_2,\ldots,x_i),x_{i+1})$$

Now let $N = (f_i)$ be a normal algorithm. We say that N is calculable on $S \subset E^\mathbf{N}$ if there exists a recursive function
$f : \mathbf{N}^2 \rightarrow \mathbf{N}$ such that :

$$\forall\, (x_n) \in S\, ,\; \forall\, i \in \mathbf{N} : f(i,\alpha_{i+1}(a^{-1}(x_0),\ldots,a^{-1}(x_i))) =$$

$$a^{-1}(f_i(x_0,x_1,\ldots,x_i))$$

This condition is equivalent to the existence of a Turing machine
which, starting with $x_0,x_1,\ldots,x_i$, stops with $f_i(x_0,x_1,\ldots,x_i)$.

On recursive functions and Turing machines, see [1],[2],[9],[34] and
[36]. The notion of decidability in the limit (the only interesting
one if we are concerned with numerical sequences) is due to Gold
([28],[29]). Lemma 1 may be given in the equivalent following form :

Lemma 1

There exists a normal algorithm satisfactory for the question "is (x_n)
convergent ?" on

  $\mathbf{U}$stati$([0,1]_\mathbf{Q}) \cup (\mathbf{AP}er[0,1]_\mathbf{Q}) -$ Conv$([0,1]_\mathbf{Q}))$.

The normal algorithm given in the proof is obviously calculable, hence
lemma 1 can be translated as :

Lemma 1'

There exists a normal calculable algorithm satisfactory for the question "is (x_n) convergent ?" on :

  $\mathbf{U}$sati$([0,1]_\mathbf{Q}) \cup (\mathbf{AP}er([0,1]_\mathbf{Q}) - \mathbf{CV}onv([0,1]_\mathbf{Q}))$

Lemma 2 used in the proof of theorem 2 may be given in the equivalent
following form :

Lemma 2

There is no normal algorithm satisfactory for the question "is (x_n)
convergent ?" on

$\text{Conv}([0,1]_Q) \cup \text{UPer}_p([0,1]_Q)$ for every $p \geq 2$.

Every normal calculable algorithm is obviously a normal algorithm !
Hence :

Lemma 2'

There is no normal calculable algorithm satisfactory for the question
"is (x_n) convergent ?" on :

$\text{Conv}([0,1]_Q) \cup \text{UPer}_p([0,1]_Q)$ for every $p \geq 2$.

This result may be further improved. Let C be the set of calculable
rational sequences of $[0,1]$. A careful study of the proof of lemma
2 shows that :

(a)   if   we   choose $\alpha_i = 1/2^i$;   then   the   sequences
$(x^0{}_n),(x^1{}_n),\ldots,(x^i{}_n),\ldots,$ are calculable,

(b) if the normal algorithm used in the proof is assumed to be calculable, then (x_n) is also calculable.

Hence :

Lemma 2"

There is no normal calculable algorithm satisfactory for the question
"is (x_n) convergent ?" on

$C \cap (\text{Conv}([0,1]_Q \cup \text{UPer}_p([0,1]_Q))$ for every $p \geq 2$.

APPENDIX 3

Decidability of the convergence, turbulence and asymptotic periodicity of a continuous function

As with the notion of algorithms for sequences, we may define the
notion of algorithm for numerical functions. Then we may consider the
problem of the algorithmic determination of the nature of a continuous
function from $[0,1]$ into itself. We give a few results.

Let E be a metric space and R be a set.

Definitions

We call an algorithm for functions from   E   into   E   (using only
functional evaluations), any sequence $A = (a_i,b_i)_{i \in \mathbf{N}}$ satisfying
$a_i : E^{2i} \to E$, $b_i : E^{2i+2} \to R$.

We call the following sequences (r_i) and (x_i) the sequence of answers given by A for f and the sequence of points used by A for f respectively :

$$x_0 = a_0 \qquad\qquad r_0 = b_0(x_0, f(x_0))$$

$$x_1 = a_1(x_0, f(x_0)) \qquad r_1 = b_1(x_0, f(x_0), x_1, f(x_1))$$

$$x_2 = a_2(x_0, f(x_0), x_1, f(x_1)) \qquad r_2 = b_2(x_0, f(x_0), x_1, f(x_1), x_2, f(x_2))$$

etc...

We say that the algorithm A is satisfactory for Q on E (a family of functions from E into E), if, for every $f \in E$, the sequence of answers given by A for f is correct for n sufficiently large. Then we say that Q is decidable in the limit on F .

Recent papers about such functions ([3],[4],[6],[7],[8],[14],[15], [16],[18],[32],[33] and [35]) lead us to define the following sets of functions from [0,1] into itself :

1) the subset of functions (called convergent) such that : for every $x_0 \in [0,1]$ the sequence $x_{n+1} = f(x_n)$ is convergent,

2) the subset of functions (called asymptotically periodic) such that, for every $x_0 \in [0,1]$, the sequence $x_{n+1} = f(x_n)$ is asymptotically periodic,

2) the subset of functions (called turbulent) such that there exists $x_0 \in [0,1]$ for which the sequence $x_{n+1} = f(x_n)$ is turbulent.

The theorems that give relationships between the nature of f and the cycles of f ([8],[16],[17] and [35]) lead one to believe that there exist sets (perhaps large) for which the question of the nature of their functions is decidable in the limit.

Fateman ([25]) has written a program which decides for certain rational functions if they are convergent or not. Unfortunately, the exact domain of efficiency of his program has not been well determined.

Using his idea, we obtain :

Proposition 1

The question "is f convergent ?" is decidable in the limit on $\mathbf{Q}[x]$ (the set of polynomial function with rational coefficients).

Sketch of the proof

At stage n , we seek the only polynomial P_n of degree $\leq n$ which matches f on (n+1) points (for example $1, 1/2, \ldots, 1/(n+1)$).

Then we divide $(P_n(X) + X)n^2$ by $P_n \circ P_n(X) - X$. If there is no remainder, we answer YES, otherwise we answer NO.

It is clear that there is no remainder if and only if $P_n(X)$ has no cycle of order 2, that is, if and only if the function $P_n(x)$ is convergent ([3],[5],[6]).

Since, for n large enough : $P_n(x) = f(x)$, the given answer is correct for n large enough.

Remark

If we assume that $f \in \mathbf{Q}[x]$ is given with its coefficients, it is clear that the proposed algorithm allows one to decide the nature of f in a finite time (and not only in the limit).

Proposition 2

The question "is f turbulent ?" is undecidable in the limit on $\mathbf{C}([0,1])$ (the set of all continuous functions from $[0,1]$ into itself).

Sketch of the proof

We use the function defined in ([16],[17]) which is turbulent, and such that :

$\forall\, n \in \mathbf{N}$, $\forall\, x_1, x_2, \ldots, x_n \in [0,1]$ $\nexists$ f asymptotically periodic such that $f(x_1) = g(x_1)$ and ... and $f(x_n) = g(x_n)$.

Remark

Many other results can be obtained in this way. Particular theorems such as the "snail theorem" ([7]) certainly give positive results of decidability (in the limit, or not).

REFERENCES

[1]   ABERT O.
"Analysis in the computable number field"
J.A.C.M., 15, 1968, pp. 275-299.

[2]   AZRA J.P. and JAULIN B.
"Récursivité"
GAUTHIER-VILLARS, Paris, 1973.

[3]   BASHUROV V.V. and OGIBIN V.N.
"Conditions for the convergence of iterative processes
on the real axis"
U.S.S.R. Computational Math. and Math. Phys., 6,5, 1966
pp. 178-184.

[4]   BUTLER G.J. and PIANIGIANI G.
"Periodic points and chaotic functions in the unit interval"
Bull. Australian Math. Soc.,8, 1978.

[5]   CHU S.E. and MOYER R.D.
"On continuous functions, commuting functions and fixed points"
Fundamenta Mathematicae LIX 1966, pp. 91-95.

[6]   COPPEL W.A.
"The solution of equations by iteration"
Trans. Cambridge Phil. Soc., 51, 1955, pp. 41-43.

[7]   COSNARD M.Y.
"On the behaviour of successive approximations"
S.I.A.M. J. Numer. Anal., 16, 1979, pp. 300-310.

[8]   COSNARD M.Y. and EBERHARD A.
"Sur les cycles d'une application continue de la variable
réelle"
Séminaire d'Analyse Numérique de Grenoble, n° 274, 1977.

[9]   DAVIS M.
"COMPUTABILITY AND UNSOLVABILITY"
Mc Graw Hill, New-York, 1959.

[10]   DELAHAYE J.P.
"Quelques problèmes posés par les suites de points non
convergentes et algorithmes pour traiter de telles suites"
Thèse de 3ème Cycle, Lille, 1979.

[11]   DELAHAYE J.P.
"Algorithmes-questions et algorithmes d'extraction pour
suites non convergentes"
Bulletin de la Direction des Etudes et Recherches de l'E.D.F.,
C, 1, 1979, pp. 17-34.

[12]   DELAHAYE J.P.
"Expériences numériques sur les algorithmes d'extraction pour
suites non convergentes"
Publication ANO n° 5, Université des Sciences et Techniques de
Lille, 1979.

[13]   DELAHAYE J.P.
"Algorithmes pour suites non convergentes"
Numer. Math., 34, 1980, pp. 333-347.

[14]   DELAHAYE J.P.
"Détermination de la période d'une suite pseudo-périodique
Bulletin de la Direction des Etudes et Recherches de l'E.D.F.,
C, 1, 1980, pp. 65-80.

[15]   DELAHAYE J.P.
"A counterexample concerning iteratively generated sequences"
J.M.A.A., 75, 1980, pp. 236-241.

[16]   DELAHAYE J.P.
"Cycles d'ordre 2^i et convergence cyclique de la méthode des
approximations successives"
Publication ANO n° 21, Université des Sciences et Techniques de
Lille, 1980.

[17]   DELAHAYE J.P.Q
"Functions admettant des cycles d'ordre n'importe quelle
puissance de 2 et aucun autre cycle"
C.R. Acad. Sc. Paris, 291, A, 1980, pp. 323-325.

[18]   DELAHAYE J.P.
"The set of periodic points"
The Amer. Math. Monthly, 88, 9, 1981, pp. 646-651.

[19]   DELAHAYE J.P.
"Décidabilité et indécidabilité à la limite de certaine
problèmes de suites"
Séminaire d'Anbalyse Numérique de Grenoble, n° 360, 1981.

[20]   DELAHAYE J.P.
"Algorithmes pour extraire une sous-suite convergente d'une
suite non convergente"
Proceeding of "Conference optimization : theory and algo-
rithms".
Lecture notes in Pure and Applied Mathematics, 1983.
Marcel Dekker.

[21]   DELAHAYE J.P.
"The cluster point set of a non convergent sequence"
Publications ANO n° 68, Université des Sciences et Techniques
de Lille, 1982.

[22]   DELAHAYE J.P. and GERMAIN-BONNE B.
        "Résultats négatifs en accélération de la convergence"
        Numer. Math., 35, 1980, pp. 443-457.

[23]   DENEL J.
        "Extensions of the continuity of point-to-set maps :
        applications to fixed point algorithms",
        Mathematical Programming Study 10, 1979, pp. 48-68.

[24]   EAVES B.C.
        "Computing Kakutani Fixed Points"
        SIAM J. Appl. Math.,21, 1971, pp. 236-244.

[25]   FATEMAN R.J.
        "An algorithm for deciding the convergence of the rational
        iteration $x_{n+1} = f(x_n)$"
        ACM Trans. on Math. Soft. 3, 1977, pp. 272-278.

[26]   FIOROT J.C. and HUARD P.
        "Composition and union of general algorithms of optimization"
        Mathematical Programming Study, 10, 1979, pp. 69-85.

[27]   GASTINEL N.
        "Introduction à l'analyse calculable"
        Cours polycopié de D.E.A., Grenoble, 1973.

[28]   GOLD E.M.
        "Limiting recursion"
        The Journal of Symbolic Logic., 30, 1965, pp. 28-48.

[29]   GOLD E.M.
        "Language identification in the limit"
        Information and Control 10, 1967, pp. 447-474.

[30]   HUARD P.
        "Optimisation dans $\mathbf{R}^n$"
        Cours de D.E.A. polycopié, Université des Sciences et
        Techniques de Lille, 1972.

[31]   HUARD P.
        "Extensions of Zangwill's Theorem"
        Mathematical Programming Study, 10, 1979, pp. 98-103.

[32]   KLOEDEN P.E.
        "Chaotic difference equations are dense"
        Bull. Austral. Math. Soc., 15, 1976, pp. 371-379.

[33]   LI T.Y. and YORKE J.A.
        "Period three implies chaos"
        The Ameri. Math., Monthly 82, 1975, pp. 985-992.

[34]   ROGERS H.
"Recursive Functions and Effective Computability"
Mc Graw Hill, New York, 1967.

[35]   SHARKOWSKII A.N.
"Co-existence of the cycles of a continuous mapping
of the line into itself"
Ukrain. M.Z., 16, 1, 1964, pp. 61,71.

[36]   TURING A.M.
"On computable numbers, with an application to the
entscheidunsproblem"
Proc. London Math. Soc. 42, 1936-7, pp. 230-265.

[37]   ZANGWILL W.I.
"Nonlinear programming : A unified approach"
Prentice Hall, Englewood Cliffs, 1969.

Chapter 3

Algorithms for Extracting
Convergent Subsequences

INTRODUCTION

Frequently in numerical analysis, we have algorithms which generate infinite sequences which are not necessarily convergent, but have interesting accumulation points. This is the case in optimization ([12],[17],[18],[23],[29],...) and with the case of iterations $x_{n+1} = f(x_n)$ (f continuous from a locally compact metric space into itself) that give rise to sequences such that $x = f^p(x)$ for every accumulation point, when there are p accumulation points ([16],[21],[22],[26],[28],...). This is also the case with the algorithms for fixed points of point-to-set maps ([7],[13],[14],[15],[19],[20],[24],[25],[27],...).

These examples motivate our study of algorithms for extracting convergent subsequences from a non-convergent sequence.

Two methods may be used :

1) Try to extract one subsequence.

2) Try to extract several subsequences, if possible converging to different accumumation points.

The first idea is studied in sections 1 and 2, the second idea in section 3.

Algorithms and results about them are given. The hypothesis made in order to obtain the convergence of the extracted subsequences are sometimes rather sharp. We show in section 4 the reason is that it is impossible to obtain extraction algorithms which are efficient on large families of sequences.

The results of this chapter have been previously presented in [3],[4],[6],[8] .

Definitions and notation

E   :   a closed subset of R^n ;

$d(x,y)$:   the distance between two points of E
              (it is not supposed that d is the Euclidian distance)

$d(A,B)$ = $\inf\{d(x,y) \mid x \in A , y \in B\}$;

$$B(x,r) \ = \ \{y \in \mathbb{R}^n | d(y,x) < r\} \ ;$$

$$\delta(A,B) \ = \ \max\{\sup\{d(A,y) | y \in B\} \ , \ \sup\{d(x,B) | x \in A\}\}$$
(Hausdorff distance between two sets; see [1]);

E^N : the set of (infinite) sequences of points of E ;

$A(x_n)$: the set of accumulation points of $(x_n) \in E^N$;
$A(x_n)$ is always closed ([1],[3],[9]) and since E
has a denumerable base of neighborhoods, one can show
that ([1],[3],[9]) :
$x \in A(x_n) \iff$ there exist subsequences of (x_n)
converging to x .

When (x_n) is given, we shall write :

$$x^p_n = (x_n, x_{n+1}, \ldots, x_p) \qquad\qquad (n \leq p)$$

For every sequence of subsets, $(A_n)_{n \in N}$ with $A_n \subset E$, we set :

$$\liminf_{n \to \infty} A_n = \{\overline{x} \in E | \forall \ \varepsilon > 0, \ \exists \ n_0, \ \forall \ n \geq n_0, \ \exists \ x \in A_n :$$

$$: \ d(\overline{x},x) \leq \varepsilon\}$$

$$\limsup_{n \to \infty} A_n = \{\overline{x} \in E | \forall \ \varepsilon > 0, \ \forall \ n_0, \ \exists \ n \geq n_0, \ \exists \ x \in A_n :$$

$$: \ d(\overline{x},x) \leq \varepsilon\}$$

If $\liminf\limits_{n \to \infty} A_n = \limsup\limits_{n \to \infty} A_n = A$,

we say that (A_n) converges to A and we denote this by :

$$\lim_{n \to \infty} A_n = A \qquad (\text{see } [1],[10],[11])$$

1 - T-ALGORITHMS

Based on a particularly simple idea, T-algorithms (presented here)
have convergence properties which are analysed on theorem 1, 2 and 3.

In particular, if $\alpha(n)$ and $\beta(n)$ are suitably chosen, it is
possible to extract convergent subsequences even if the sequence to be
treated has an infinite accumulation point set; with S and
U-algorithms this is no longer possible.

In addition to the sequence (x_n) of points of E , throughout this
section, let $\alpha(n)$, $\beta(n)$ be two integer sequences with the following
properties :

$$(\alpha\beta) \quad \left| \begin{array}{l} \alpha(0) = 0 \\[2mm] \forall\ n \in \mathbf{N} : \alpha(n+1) > \alpha(n) \\ \qquad\qquad \beta(n) \geq \alpha(n+1)-1 \end{array} \right.$$

Of course $(\alpha\beta)$ implies :

$$\lim_{n\to\infty} \alpha(n) = \lim_{n\to\infty} \beta(n) = +\infty \quad \text{and} \quad \forall\ n \in \mathbf{N} \quad \beta(n) \geq \alpha(n)$$

Sequences $(\alpha(n),\beta(n))$ permit us to cut the sequence (x_n) into "slices" :

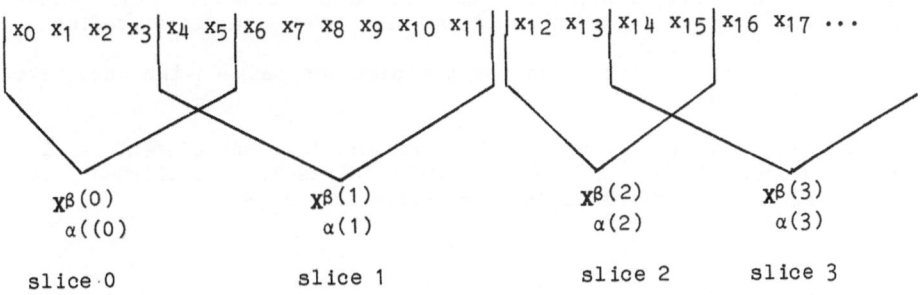

Figure 1

T-algorithms extract a subsequence by taking one point from each slice.

Let i_0 be a fixed integer.

Algorithm $T(i_0,\alpha(p),\beta(p))$

Step 0

Set $t_0 = x_{i_0}$

Let p_0 be the least integer satisfying :

$$\alpha(p_0) \leq i_0 \leq \beta(p_0)$$

Step j

Let I_j be the set of integers ℓ such that :

$$\alpha(p_0+j) \leq \ell \leq \beta(p_0+j) ,$$

$$d(x_\ell, t_{j-1}) = \min\{d(x_m, t_{j-1}) \mid \alpha(p_0+j) \leq m \leq \beta(p_0+j)\} .$$

set $i_j = \max I_j$

set $t_j = x_{i_j}$

When parameters $i_0, (\alpha(n), \beta(n))$ are fixed, for every sequence (x_n) the algorithm $T(i_0, \alpha(n), \beta(n))$ extracts a subsequence (t_j) which begins with $t_0 = x_{i_0}$ and then takes point from each slice, defined by $(\alpha(n), \beta(n))$, which minimizes the distance between two succesive extracted points.

We set $i_j = \max I_j$, in case there is more than one element in I_j, but this choice is not really essential and results established later are still true if one makes another choice for $i_j \in I_j$.

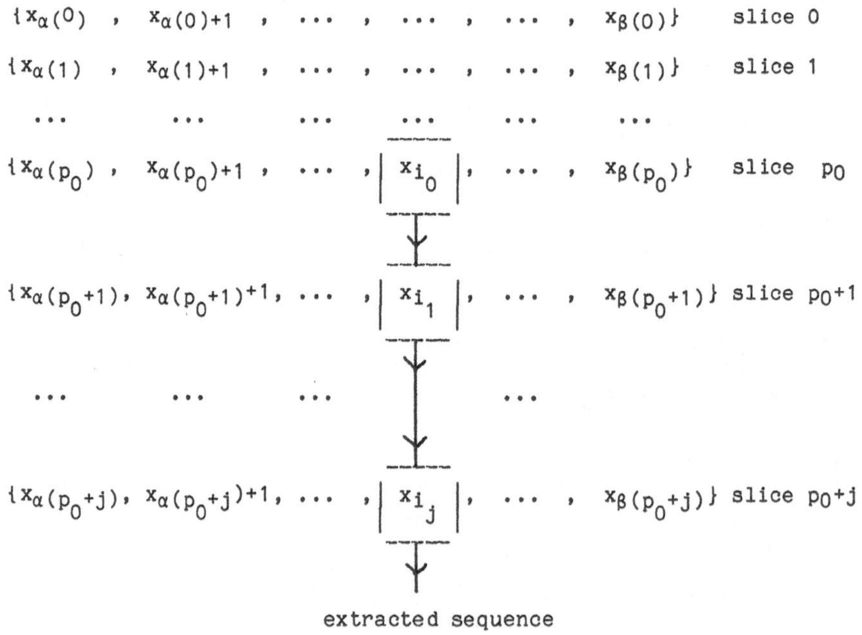

$$\{x_{\alpha(0)} , \quad x_{\alpha(0)+1} , \quad \cdots , \quad \cdots , \quad \cdots , \quad x_{\beta(0)}\} \qquad \text{slice 0}$$

$$\{x_{\alpha(1)} , \quad x_{\alpha(1)+1} , \quad \cdots , \quad \cdots , \quad \cdots , \quad x_{\beta(1)}\} \qquad \text{slice 1}$$

$$\cdots \qquad \cdots \qquad \cdots \qquad \cdots \qquad \cdots \qquad \cdots$$

$$\{x_{\alpha(p_0)} , \quad x_{\alpha(p_0)+1} , \quad \cdots , \quad \boxed{x_{i_0}} , \quad \cdots , \quad x_{\beta(p_0)}\} \qquad \text{slice } p_0$$

$$\{x_{\alpha(p_0+1)} , x_{\alpha(p_0+1)+1}, \cdots , \boxed{x_{i_1}} , \cdots , x_{\beta(p_0+1)}\} \text{ slice } p_0+1$$

$$\cdots \qquad \cdots \qquad \cdots \qquad \cdots$$

$$\{x_{\alpha(p_0+j)}, x_{\alpha(p_0+j)+1}, \cdots , \boxed{x_{i_j}} , \cdots , x_{\beta(p_0+j)}\} \text{ slice } p_0+j$$

extracted sequence

Figure 2

If the subsequence $(x_{i_j}) = (t_j)$ converges to x, we write :

$$(i_0 , \alpha(n) , \beta(n)) \rightarrow x$$

or, for simplicity : $i_0 \rightarrow x$

To indicate that $(x_{i_j}) = (t_j)$ is convergent, we write :

$$(i_0 , \alpha(n) , \beta(n)) \downarrow$$

or, for simplicity : $i_0 \downarrow$

Example 1

$E = \mathbf{R}$. Let (x_n) be the sequence defined by :

$$x_{3n} = 1/2^n$$

$$x_{3n+1} = 1 + 1/2^n$$

$$x_{3n+2} = 2 + 1/2^n$$

(a) $\alpha(n) = 3n$, $\beta(n) = 3n+2$.

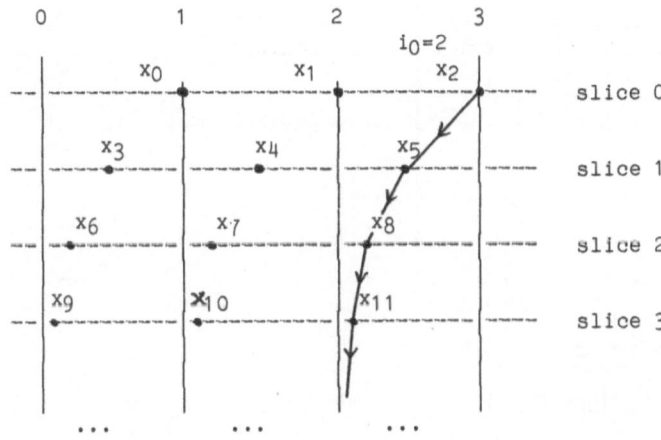

Figure 3

If $i_0 = 0$, we obtain the subsequence $(x_0, x_4, x_7, x_{10}, \ldots)$ which converges to $1 \in \mathbf{A}(x_n)$. This is denoted by :

$$i_0 = 0 \downarrow 1$$

Similarly :
$$i_0 = 1 \downarrow 2$$

$$i_0 = 2 \downarrow 2$$

$$i_0 = 3 \downarrow 0$$

In this case, we obviously obtain that :

(*) $$\forall\, i_0 \in \mathbf{N} : i_0 \downarrow$$

and

(**) $$\forall\, x \in A(x_n) \,,\, \exists\, i_0 \in \mathbf{N} : i_0 \downarrow x \ .$$

(b) $\alpha(n) = 4n$, $\beta(n) = 4n+3$.

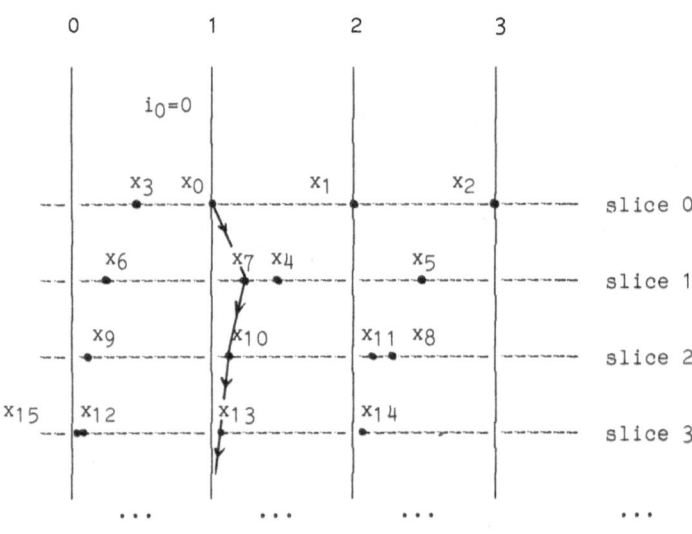

Figure 4

This new sequences give us :

$$i_0 = 0 \downarrow 1$$

$$i_0 = 1 \downarrow 2$$

$$i_0 = 2 \downarrow 2$$

$$i_0 = 3 \downarrow 0$$

Properties (*) and (**) are still true.

More generally, it is easy to show that they are true if :

$$\forall\ n \in \mathbf{N} :\ \alpha(n+1) \geq \alpha(n) + 3$$

Example 2

$E = \mathbf{R}$. We define (x_n) by :

$$(x_n) = (0,0,1/2,0,1/4,1/2,3/4,0,1/8,1/4,3/8,1/2,5/8,3/4,7/8,\ldots)$$

We immediatly obtain :

$$A(x_n) = [0,1]$$

In order to obtain convergent subsequences with T , it is necessary to choose longer and longer slices. For instance, we can set :

$$\alpha(p) = 2^p-1\ ,\quad \beta(p) = 2^{p+1}-2.$$

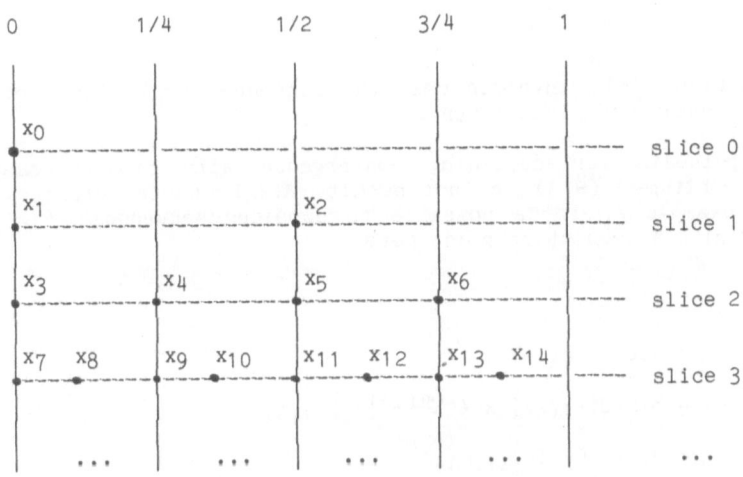

Figure 5

Property (*) holds, but not (**). The points $x \in A(x_n)$ such that $\nexists\ i_0 : i_0 \to x$ may be characterized : they are points $k/2^i$ with $k \in \mathbf{N}$, $i \in \mathbf{N}$, $0 \leq k \leq 2^i-1$. Since the possible i_0 are denumerable, the set of points which are the limit of some subsequence extracted by T is at most denumerable. This is still true, even if we define slices to be even longer. We remark that with the sequence of example 2, if $\beta(n) - \alpha(n)$ is bounded, then whatever the value of i , the sequence $(x_{i_j}) = (t_j)$ does not converge (in fact we have $A(t_j) = [0,1]$). Hence, for such sequences it is absolutely necessary to define long slices.

The previous examples show that the   T   algorithms are not always convergent, and even if they converge they do not give every point of $A(x_n)$.   To obtain convergent subsequences, we have to make good choices for $(\alpha(n),\beta(n))$.

Theorem 1

If the following condition is satisfied :

$$(\sigma) \quad \sum_{i \in \mathbf{N}} \delta(X^{\beta(i)}_{\alpha(i)} \quad , \quad X^{\beta(i+1)}_{\alpha(i+1)}) < \infty$$

then :

(i)   for every $i_0 \in \mathbf{N}$: $i_0 \downarrow$ (i.e.   T   gives only convergent subsequences)

(ii)   for every isolated point   x   of $A(x_n)$ there is $i_0 \in \mathbf{N}$ such that $i_0 \downarrow x$.

Remarks

1)   The condition (σ) involves only the sequence (x_n) (and not $A(x_n)$ as the conditions given later).

2)   The hypothesis for obtaining convergence with simultaneous extraction algorithms   (§ 3)   do not permit $A(x_n)$ to be infinite. Here, as in example 2, it is possible to consider sequences (x_n) having an infinite accumulation point sets.

Proof

(i)   From the definition of   T :

$$d(t_i, t_{i+1}) = \min\{d(t_i, x) \mid x \in X^{\beta(i+1)}_{\alpha(i+1)}\}$$

$$= d(t_i, X^{\beta(i+1)}_{\alpha(i+1)})$$

Hence :

$$d(t_i, t_{i+1}) \leq \sup\{d(y, X^{\beta(i+1)}_{\alpha(i+1)} \mid y \in X^{\beta(i)}_{\alpha(i)}\}$$

$$\leq \delta(X^{\beta(i)}_{\alpha(i)}, X^{\beta(i+1)}_{\alpha(i+1)}).$$

This gives $\sum_{i \in \mathbf{N}} d(t_i, t_{i+1}) < \infty$.

Consequently, (t_i) is a Cauchy sequence, and thus converges.

(ii) Let $\varepsilon = d(x, A(x_n) - \{x\}) > 0$.

There exists i_0 such that :

$$\sum_{i \geq i_0} \delta(X^{\beta(i)}{}_{\alpha(i)}, X^{\beta(i+1)}{}_{\alpha(i+1)}) < \varepsilon/3 \quad \text{and} \quad x_{i_0} \in B(x, \varepsilon/3)$$

From (i), the sequence (t_i) converges to $y \in A(x_n)$.
y necessarily satisfies :

$$d(x_{i_0}, y) < 2 \ \varepsilon/3$$

Hence $y = x$.

Theorem 1 leads directly to the question : "is it always possible to find sequences $\alpha(n)$, $\beta(n)$ satisfying (σ) (and $(\alpha\beta)$) ?"

The following proposition answers this question :

Proposition 1

Let (x_n) be a bounded sequence of points in E . Then, there exist sequences $\alpha(n)$, $\beta(n)$ such that (σ) and $(\alpha\beta)$ are satisfied.

Proof

For each $\varepsilon > 0$, we define :

$$V_\varepsilon = \{x \mid d(x, A(x_n)) < \varepsilon\}$$

Since (x_n) is a bounded sequence in E , $A(x_n)$ is compact and thus :

$$\forall \ \varepsilon > 0 \ , \ \exists \ n(\varepsilon) \ , \ \forall \ m \geq n(\varepsilon) : x_m \in V_\varepsilon.$$

Let $(\varepsilon_i)_{i \geq 1}$ be a sequence of parameters such that :

$$\varepsilon_i > 0 \ , \ \sum_{i \geq 1} \varepsilon_i < \infty \ .$$

We set : $\alpha(0) = 0$, $\beta(0) = n(\varepsilon_1)$, $\alpha(1) = \beta(0) + 1$.

It is possible to cover $A(x_n)$ with a finite number of balls of radius $\varepsilon_1/2$:

$$\left|\begin{array}{l} B(y^1{}_0, \varepsilon_1/2) \ldots, B(y^1{}_{\ell_1}, \varepsilon_1/2) \\[2mm] y^1{}_0, \ldots, y^1{}_{\ell_1} \in A(x_n) \end{array}\right.$$

We take k_1 such that for every $i \in \{0,\ldots,\ell_1\}$, there exists $n \in \{\alpha(1),\ldots,k_1\}$ satisfying $(x_n) \in B(y^1{}_i,\varepsilon_1/2)$. Then we set :

$$\beta(1) = \max\{k_1,\alpha(1),n(\varepsilon_2)\} \ , \ \alpha(2) = \beta(1) + 1.$$

We now establish that :

(*) $$\delta(X^{\beta(1)}{}_{\alpha(1)},\mathbf{A}(x_n)) \leq \varepsilon_1 \ .$$

Let $x \in X^{\beta(1)}{}_{\alpha(1)}$. Since $\alpha(1) \geq n(\varepsilon_1)$, we have that $d(x,\mathbf{A}(x_n)) \leq \varepsilon_1$.

Let $\bar{x} \in \mathbf{A}(x_n)$. There exists $y^1{}_r$ such that :

$$d(\bar{x},y^1{}_r) \leq \varepsilon_1/2 \ .$$

Hence, there exists $n \in \{\alpha(1),\ldots,\beta(1)\}$ such that :

$$d(y^1{}_r,x_n) \leq \varepsilon_1/2 \ ;$$

which implies that $d(\bar{x},x_n) \leq \varepsilon_1$.

This means that $d(\bar{x},X^{\beta(1)}{}_{\alpha(1)}) \leq \varepsilon_1$, and finally gives (*).

Now, we cover $\mathbf{A}(x_n)$ with a finite number of balls of radius $\varepsilon_2/2$:

$$\left|\begin{array}{l} B(y^2{}_0,\varepsilon_2/2,\ldots,B(y^2{}_{\ell_2},\varepsilon_2/2) \\ y^2{}_0,\ldots,y^2{}_{\ell_2} \in \mathbf{A}(x_n) \end{array}\right.$$

We take k_2 such that, for every $i \in \{0,1,\ldots,\ell_2\}$, there exists $n \in \{\alpha(2),\ldots,k_2\}$ satisfying $(x_n) \in B(y^2{}_i,\varepsilon_2/2)$. Then we set :

$$\beta(2) = \max\{k_2,\alpha(2),n(\varepsilon_2)\} \ , \ \alpha(3) = \beta(2) + 1 \ ,$$

and establish that :

$$\delta(X^{\beta(2)}{}_{\alpha(2)} \ , \ \mathbf{A}(x_n)) \leq \varepsilon_2 \ .$$

We continue this construction analogously, and finally obtain that :

$$\sum_{i \in \mathbb{N}} \delta(X^{\beta(i)}{}_{\alpha(i)},X^{\beta(i+1)}{}_{\alpha(i+1)}) \leq 2 \sum_{i \in \mathbb{N}} \delta(X^{\beta(i)}{}_{\alpha(i)},\mathbf{A}(x_n)) < \infty \ .$$

Remarks

1) In fact the proof establishes the following more precise result :

   if (x_n) is bounded and if $(\varepsilon_n)_{n \geq 1}$ is a sequence of positive real numbers then there exists $\alpha(n)$ such that :
   $\forall \ m \geq 1 \quad \delta(X^{\alpha(m+1)-1}{}_{\alpha(m)} \ , \ \mathbf{A}(x_n)) \leq \varepsilon_m \ .$

This means that the accumulation point set of a sequence can be arbitrarily well approximated, using only complementary slices of (x_n). These results are related to some of $[3]$.

2) If, instead of (σ) , we simply suppose that :

(σ') $\displaystyle\lim_{i \to \infty} \delta(X^{\beta(i)}_{\alpha(i)} , X^{\beta(i+1)}_{\alpha(i+1)}) = 0$

or

(σ'') $\displaystyle\lim_{i \to \infty} X^{\beta(i)}_{\alpha(i)} = A(x_n)$,

then, theorem 1 does not remain true. This is shown by the following counterexample :

Counterexample

Consider the following sequence (see Fig. 6) :

$(x_n) = (0,1/4,1/2,3/4,0,1/2,1,3/8,7/8,1/4,3/4,...)$.

With $\alpha(n) = (0,1,2,3,4,7,9,11,13,...)$ (see Fig. 6), and

$\beta(n) = \alpha(n+1)-1$.

We obtain : $A(x_n) = [0,1]$

Conditions (σ') and (σ'') are satisfied, but not condition (σ).

It is obvious that, for every $i_0 \in \mathbf{N}$, the sequence (x_{i_j}) does not converge and satisfies : $A(x_{i_j}) = [0,1]$.

Theorem 2

If the sequence (x_n) is bounded, has only a finite number of accumulation points, and satisfies one of the following conditions :

(σ') $\displaystyle\lim_{i \to +\infty}$ $\delta(X^{\beta(i)}_{\alpha(i)} , X^{\beta(i+1)}_{\alpha(i+1)}) = 0$

(σ'') $\displaystyle\lim_{i \to \infty}$ $X^{\beta(i)}_{\alpha(i)} = A(x_n)$

then :

(i) for every $i_0 \in \mathbf{N}$: $i_0 \downarrow$

(ii) for every $x \in A(x_n)$ there exists $i_0 \in \mathbf{N}$ such that :
     $i_0 \downarrow x$.

Remark

$(\sigma) \Rightarrow (\sigma')$; $(\sigma) \Rightarrow (\sigma'')$, thus hypothesis (σ') and (σ'') are more often satisfied than (σ) , and, from proposition 1, we can say that, for a given (x_n) , it is always possible to find $\alpha(n)$ and $\beta(n)$ satisfying (σ') , (σ'') .

Proof

Let the accumulation points of (x_n) be denoted by $y_1, y_2, \ldots, y_\ell$.

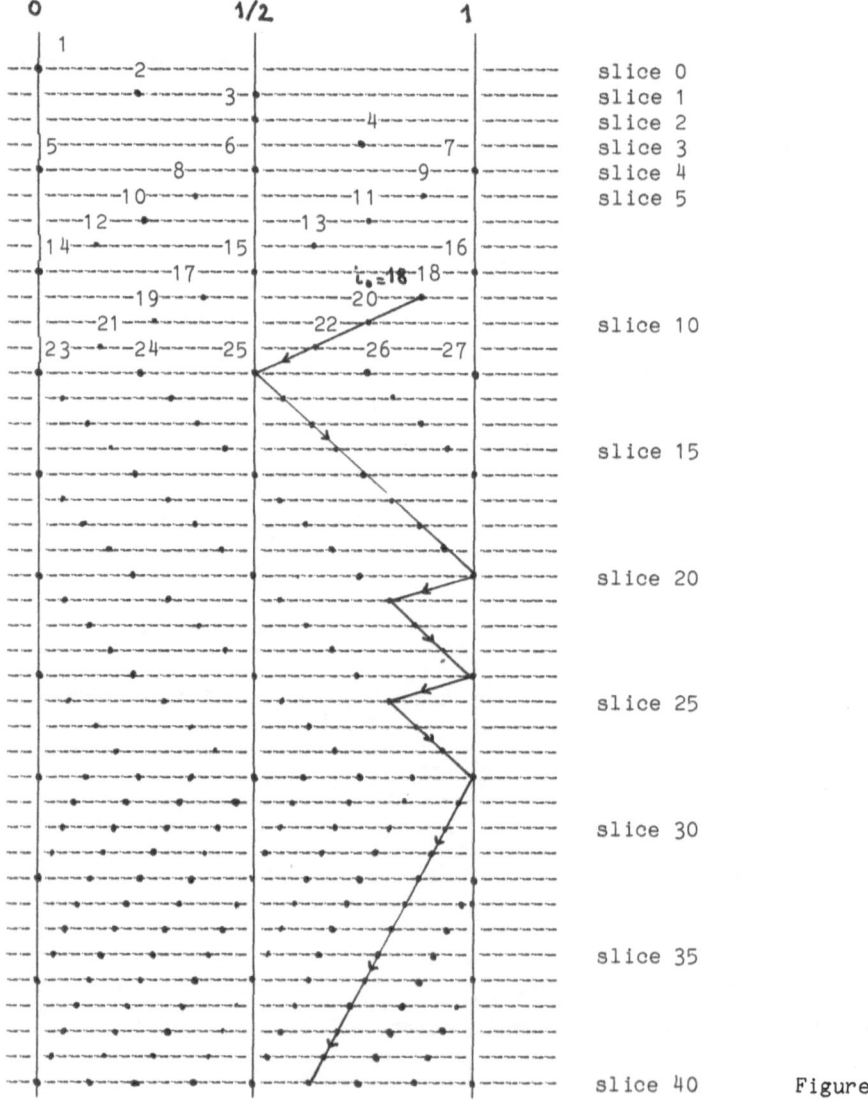

Figure 6

Let $\varepsilon_0 = \min\{d(y_i, y_j) ; i \neq j\}$

Using the fact that (x_n) is bounded and has a finite number of accumulation points, we shall see that $(\sigma') \iff (\sigma'')$.

a) $(\sigma') \Rightarrow (\sigma'')$.

Let i_0 be an integer such that :

$$\forall\, i \geq i_0 : \delta(X^{\beta(i)}{}_{\alpha(i)} , X^{\beta(i+1)}{}_{\alpha(i+1)}) \leq \varepsilon_0/4 ,$$

$$\forall\, m \geq \alpha(i_0) : d(x_m, A(x_n)) \leq \varepsilon_0/4 .$$

Reasoning by contradiction, we obtain that :

$$\forall\, i \geq i_0 , \forall\, j \in \{1, 2, \ldots, \ell\} : B(y_j, \varepsilon/4) \cap X^{\beta(i)}{}_{\alpha(i)} \neq \emptyset$$

which gives us :

$$\liminf_{i \to \infty} X^{\beta(i)}{}_{\alpha(i)} \supset A(x_n)$$

Since the inclusion :

$$\liminf_{i \to \infty} X^{\beta(i)}{}_{\alpha(i)} \subset A(x_n)$$

is always satisfied, we obtain $(\sigma') \Rightarrow (\sigma'')$.

b) $(\sigma'') \Rightarrow (\sigma')$.

For every $r \in \{1, \ldots, \ell\}$, let $(x_{k_r}(i))$ be a subsequence converging to y_r and such that $k_r(i) \in \{\alpha(i), \ldots, \beta(i)\}$.

For every $\varepsilon \in R^{+*}$, there exists $n(\varepsilon)$ such that :

$$\forall\, m \geq n(\varepsilon) : d(x_m, A(x_n)) \leq \varepsilon ,$$

$$\forall\, r \in \{1, \ldots, \ell\} : k_r(m) \geq n(\varepsilon) \Rightarrow d(x_{k_r}(m), y_r) \leq \varepsilon$$

Then, it is obvious that, for every i, such that $\alpha(i) \geq n(\varepsilon)$:

$$\delta(X^{\beta(i)}{}_{\alpha(i)} , X^{\beta(i+1)}{}_{\alpha(i+1)}) \leq 2\,\varepsilon ,$$

This shows (σ').

c) Part (i) of theorem 2

A reasoning similar to the proof of theorem 1 shows that for every $i_0 \in N$:

$$\lim_{i \to \infty} d(t_i, t_{i+1}) = 0$$

Thus $A(t_i)$ is connected $([2],[7])$. Since $A(t_i) \subset A(x_n)$, (t_i) converges.

d) Part (ii) of theorem 3

If $y_r \in A(x_n)$ we define i_0 as follows :

$$\alpha(i) \geq i_0 \Rightarrow \delta(X^{\beta(i)}{}_{\alpha(i)}, X^{\beta(i+1)}{}_{\alpha(i+1)}) \leq \epsilon_0/4;$$

$$m \geq i_0 \Rightarrow d(x_m, A(x_n)) \leq \epsilon_0/4 ;$$

$$d(x_{i_0}, y_r) \leq \epsilon_0/4$$

Theorem 3

(i) If (x_n) is asymptotically periodic, then with $\alpha(n) = n$ and $\beta(n) = 2n$, we have :

$$\forall~ i_0 : i_0 \downarrow ~~\text{and}$$

$$\forall~ x \in A(x_n) ~~\dashv i_0 : i_0 \downarrow x ~.$$

(ii) If x is an isolated accumulation point with a strictly positive strength, then with $\alpha(n) = n$ and $\beta(n) = n^2$, we have :

$$\dashv i_0 : i_0 \downarrow x ~.$$

Proof

(i) The hypothesis of theorem 2 are fulfilled.

(ii) Let $\epsilon = d(x, A(x_n) - \{x\}) > 0$.

There exists n_0 such that if $n \geq n_0$, then :

$$X^{n^2}_n \cap B(x, \epsilon/5) \neq \emptyset$$

and

$$x_n \in B(x, \epsilon/5) ~~\text{or}~~ x_n \notin B(x, 3\epsilon/5)$$

We choose i_0 such that $i_0 \geq n_0$ and $d(x_{i_0}, x) \leq \epsilon/5$.

2 – S-ALGORITHMS

Let $(x_n,)$, $(\alpha(n), \beta(n))$ be given as in section 1.

Let $a \in E$.

Algorithm $S(a, \alpha(n), \beta(n))$

Step j

Let I_j be the set of integers ℓ such that :

$$\alpha(j) \leq \ell \leq \beta(j)$$

$$d(x_\ell, a) = \min\{d(x_m, a) | \alpha(j) \leq m \leq \beta(j)\}.$$

Set $i_j = \max I_j$.

Set $S_j = x_{i_j}$.

When parameters $a, (\alpha(n), \beta(n))$ are fixed, for every sequence (x_n) ,
the algorithm $S(a, \alpha(n), \beta(n))$ extracts a subsequence (s_j) obtained
by taking out the closest point   a   among the points of the slice
defined by $\alpha(n), \beta(n)$ (if there are several, the point of largest
index is chosen).

If the subsequence $(x_{i_j}) = (s_j)$ converges to x   we write :

$$(a, \alpha(n), \beta(n)) \uparrow x$$

or, for simplicity : $a \uparrow x$.

To indicate that $(x_{i_j}) = (s_j)$ converges, we write :

$$(a, \alpha(n), \beta(n)) \uparrow$$

or, for simplicity : $a \uparrow$.

Example 1

$E = \mathbf{R}.$

Let (x_n) be the sequence defined by :

$$x_{2n} = - (1 + 1/2^n)$$

$$x_{2n+1} = + (1 + 1/2^n)$$

(a) $\alpha(n) = 2n$, $\beta(n) = 2n+1$

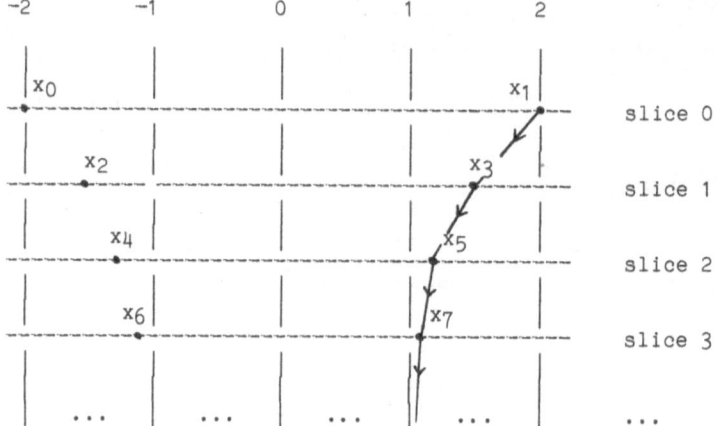

Figure 7

If we choose $a \geq 0$, we obtain : $a \uparrow 1$

If we choose $a < 0$, we obtain : $a \uparrow -1$

Thus, we have :

(*) $\forall\ a\ :\ a \uparrow$

(**) $\forall\ x \in \mathbf{A}(x_n)\ \not\exists\ a\ :\ a \uparrow x$

(b) $\alpha(n) = 3n$, $\beta(n) = 3n+2$.

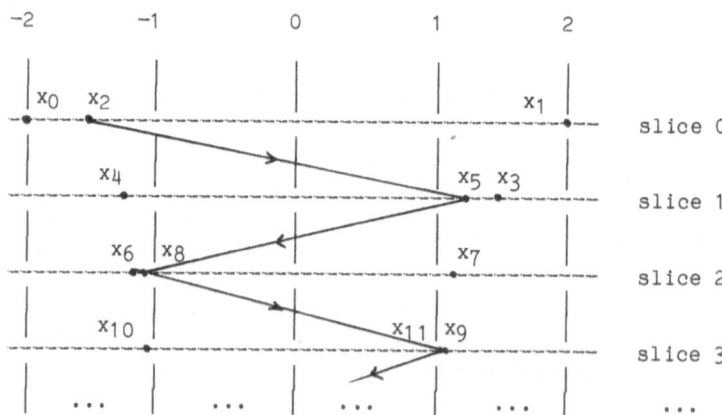

Figure 8

If we have $a > 0$, we obtain : $a \uparrow +1$

If we have $a < 0$, we obtain : $a \uparrow -1$

But, with a = 0 , we have :

$$(s_i) = (x_2, x_5, x_8, x_{11}, \ldots, x_{3n-1}, \ldots) .$$

Thus, (s_i) does not converge.

Example 2

E = **C.** (the complex field)

$$x_n = e^{in(1/2^n)}$$

$$A(x_n) = \{z \in \mathbf{C}; \ |z| = 1\}$$

Let C be a constant. If $\beta(n) - \alpha(n) < C$, then there is no a such that $a \uparrow$.

With $\alpha(n) = 2^n$, $\beta(n) = \alpha(n+1) - 1$, we obtain :

$$(a \neq 0 \ \text{ and } \ \arg(a) = \theta) \Rightarrow a \uparrow e^{i\theta}$$

$$a = 0 \Rightarrow A(s_i) = A(x_n)$$

Theorem 4

If the following conditions are satisfied :

(σ'') $\lim\limits_{i \to \infty} X\beta(i)_{\alpha(i)} = A(x_n)$

(γ) $\text{card}\{x \in A(x_n) | d(a,x) = d(a, A(x_n))\} = 1$

then : $a \uparrow$.

Remark

From proposition 1 (see also [2]), it is always possible to choose $(\alpha(n), \beta(n))$ satisfying (σ''). Similarly, it is clear that we can choose a satisfying (γ) . However, the hypothesis of theorem 4 are less convenient than these of theorem 1, for they essentially involve $A(x_n)$.

Proof

Let x be the unique element of $A(x_n)$ such that :

$$d(a,x) = d(a, A(x_n))$$

From $(\sigma")$: $x \in \liminf_{j \to \infty} X^{\beta(i)}{}_{\alpha(i)}$

Thus there is a subsequence (x_{k_j}) converging to x and such that :

$$\forall j \in \mathbf{N} : x_{k_j} \in X^{\beta(i)}{}_{\alpha(i)} \ .$$

Hence :

$$\lim_{j \to \infty} d(a, x_{k_j}) = d(a, x)$$

This relation and the fact that closed balls are compact, implies that

$$\lim_{j \to \infty} x_{k_j} = x$$

3 - U-ALGORITHMS

The U-algorithms are based on a more complicated idea that T and S-algorithms. They are devoted to giving finer results on the initial sequence (x_n) :   they have to extract simultaneously convergent subsequences.

Let k be a fixed integer (k is the number of subsequence to be extracted).   Let $\alpha(n)$, $\beta(n)$ be two integer sequences satisfying :

$$\forall p \in \mathbf{N} : \beta(p) \geq \alpha(p) + k-1 \ ,$$

$$\alpha(p) \geq p \ .$$

Algorithm U$(k,\alpha(p),\beta(p))$

Step p

Previous steps have determined k finite subsequences

$$(x^1{}_0, x^1{}_1, \ldots, x^1{}_{p-1})(x^2{}_0, x^2{}_1, \ldots, x^2{}_{p-1}), \ldots, (x^k{}_0, x^k{}_1, \ldots, x^k{}_{p-1}) \ ;$$

step p   determines k points $x^1{}_p, x^2{}_p, \ldots, x^k{}_p$ for lengthening each sequence by one point.

Search of k mutually distant points

(1) Set $a_1 = x_{\beta(p)}$,

(2) Set $A_1 = \{a_1\}$, and determine :

$$x_i , \ i \in \{\alpha(p),\alpha(p)+1,\ldots,\beta(p)\} ,$$

the most distant point of A_1 (if there are several, choose the one with largest index). Set $a_2 = x_i$.

(3) Set $A_2 = \{a_1,a_2\}$, and determine :

$$x_i, \ i \in \{\alpha(p),\alpha(p+1),\ldots,\beta(p)\},$$

the most distant point of A_2 . Set $a_3 = x_i$.

...   .

(k) Set $A_{k-1} = \{a_1,a_2,\ldots,a_{k-1}\}$, and determine :

$$x_i , \ i \in \{\alpha(p),\alpha(p+1),\ldots,\beta(p)\} ,$$

the most distant point of A_{k-1}. Set $a_k = x_i$.

Lengthening of finite subsequences

(1) x^1_p is the closest point a_i of x^1_{p-1} (if there are several choose the one with largest index).

(2) x^2_p is the closest point a_i $(i \in \{1,2,\ldots,k\} \ a_i \neq x^1_p)$ of x^2_{p-1}

...   .

(k) x^k_p is the unique remaining a_i of

$$\{a_1,a_2,\ldots,a_k\} - \{x^1_p,x^2_p,\ldots,x^{k-1}_p\} .$$

(At step 0 , the lengthening is :

$$x^1_p = a_1 , \ x^2_p = a_2,\ldots,x^k_p = a_k).$$

Example 1

Let $(y_n),(z_n),(x_n)$ be defined by :

$$(y_n) = (3,2,3,1,2,3,2,3,1,2,\ldots) ;$$

$$(z_n) = (2,-2,1,-1,1/2,-1/2,1/4,-1/4,1/8,-1/8,\ldots) ;$$

$$(x_n) = (y_n + z_n) .$$

Algorithm $U(3,5p,5p+4)$ is applied. First steps are :

Step 0   points under consideration : x_0, x_1, x_2, x_3, x_4.

$a_1 = x_4$; $a_2 = x_3$; $a_3 = x_0$;

$x^1_0 = a_1$; $x^2_0 = a_2$; $x^3_0 = a_3$.

Step 1   points under consideration : x_5, x_6, x_7, x_8, x_9.

$a_1 = x_9$; $a_2 = x_7$; $a_3 = x_8$;

$x^1_1 = a_2$; $x^2_1 = a_3$; $x^3_1 = a_1$;

Step 2   points under consideration : $x_{10}, x_{11}, x_{12}, x_{13}, x_{14}$.

$a_1 = x_{14}$; $a_2 = x_{13}$; $a_3 = x_{12}$;

$x^1_2 = a_3$; $x^2_2 = a_2$; $x^3_2 = a_1$.

It is clear that sequence $(x^1_p), (x^2_p), (x^3_p)$ converge respectively to
3 , 1 and 2 , which are the three accumulation points of (x_n).

Example 2

It is possible that some of constructed subsequences are non convergent. Here is an example :

$(y_n) = (0,0,5,0,5,5,0,0,5,0,5,5,...)$;

$(z_n) = (0,1,0,0,0,1,0,1/2,0,0,0,1/2,...)$;

$(x_n) = (y_n + z_n)$.

Algorithm $U(3, 3p, 3p+2)$ is applied. First steps are :

Step 0 $x^1_0 = x_2$, $x^2_0 = x_0$, $x^3_0 = x_1$.

Step 1 $x^1_1 = x_4$, $x^2_1 = x_3$, $x^3_1 = x_5$.

Step 2 $x^1_2 = x_8$, $x^2_2 = x_6$, $x^3_2 = x_7$.

Step 3 $x^1_3 = x_{10}$, $x^2_3 = x_9$, $x^3_3 = x_{11}$.

Step 4 $x^1_4 = x_{14}$, $x^2_4 = x_{12}$, $x^3_4 = x_{13}$.

Step 5 $x^1_5 = x_{16}$, $x^2_5 = x_{15}$, $x^3_5 = x_{17}$.

Subsequences (x^1_n) and (x^2_n) converge. Subsequence (x^3_n) does
not converge.

Theorem 5

For every bounded sequence (x_n) in $\mathbf{R}^m$ having k' accumulation points $y_1, y_2, \ldots, y_{k'}$ with strictly positive strength, algorithm $U(k, p, p^2+k)$ $k \geq k'$ gives k subsequences, at least one of them converging to y_1 (resp. $y_2, y_3, \ldots, y_{k'}$). In particular, if $k = k'$, all subsequences are convergent (to $y_1, y_2, \ldots, y_k$).

Remarks

1) Theorem 5 can be extended as follows :

(a) instead of $\mathbf{R}^m$, one can take any locally compact metric space.

(b) instead of $U(k, p, p^2+k)$, one can take $U(k, \alpha(p), \beta(p))$ with

$$\lim_{p \to \infty} \alpha(p) = + \infty$$

$$\forall\, p \in \mathbf{N} : \beta(p) \geq \alpha(p) + k ,$$

$$\forall\, \lambda > 1 , \; \exists\, C \in \mathbf{R}^{+*} , \; \exists\, p_0 \in \mathbf{N} , \; \forall\, p \geq p_0 : \beta(p) \geq C(\alpha(p))^\lambda .$$

(c) The search of k mutually distant points is possible with other algorithms. For instance, one can first try to obtain a natural partition of the set $\{x_{\alpha(p)}, x_{\alpha(p)+1}, \ldots, x_{\beta(p)}\}$ (with the help of automatic classification methods ([2],[3])), then take out one a_i from each set of the derived partition ([3]).

2) Numerical experiments made on these algorithms shows that only short execution times are necessary in order to obtain satisfactory results [4].

Proof

Let (x_n) be a sequence satisfying the assumption of the theorem.

Set : $\varepsilon = \min\, d(y_i, y_j) | i, j \in \{1, 2, \ldots, k'\}\; i \neq j\}$;

$$V = B(y_1, \varepsilon/5) \cup \ldots \cup B(y_{k'}, \varepsilon/5).$$

Let $p_0 \in \mathbf{N}$ be such that :

$$\forall\, p \geq p_0 : x_p \in V .$$

There exists $p_1 \geq p_0$ such that :

$$\forall\, p \geq p_1 , \; \forall\, i \in \{1, 2, \ldots, k'\} : \{x_p, x_{p+1}, \ldots, x_{\beta(p)}\} \cap B(y_i, \varepsilon/5) \neq \emptyset ,$$

where $\beta(p) = p^2 + k$.

For every $p \geq p_1$, the sets

$$\{x_p, x_{p+1}, \ldots, x_{\beta(p)}\} \cap B(y_i, \varepsilon/5) , i \in \{1, 2, \ldots, k'\}$$

are a partition of $\{x_p, x_{p+1}, \ldots, x_{\beta(p)}\}$.

It is obvious that at step p , the search of k mutually distant points gives k points $a_1, a_2, \ldots, a_k$ such that each ball $B(y_i, \varepsilon/5)$ contains at least one of them.

Let $i(1) \in \{1, 2, \ldots, k'\}$ be such that $x^1{}_{p_1} \in B(y_{i(1)}, \varepsilon/5)$.

By recurrence, it is clear that :

$$\forall p \geq p_1 : x^1{}_p \in B(y_{i(1)}, \varepsilon/5) ;$$

thus the subsequence $(x^1{}_p)$ converges to $y_{i(1)}$.

Let $i(2) \in \{1, 2, \ldots, k'\}$ such that $x^2{}_{p_2} \in B(y_{i(2)}, \varepsilon/5)$.

Three cases are possible :

(a) $i(2) \neq i(1)$

We show that $\forall p \geq p_1 : x^2{}_p \in B(y_{i(2)}, \varepsilon/5)$; thus the subsequence $(x^2{}_p)$ converges to $y_{i(2)}$.

(b) $i(2) = i(1)$ and for every $p \geq p_1$:

there are at least two points a_i ($i \in \{1, 2, \ldots, k\}$) in the ball $B(y_{i(p)}, \varepsilon/5)$.

We show that : $\forall p \geq p_1$ $x^2{}_p \in B(y_{i(p)}, \varepsilon/5)$; thus the subsequence $(x^2{}_p)$ converges to $y_{i(1)}$.

(c) $i(2) = i(1)$ and there exists $p \geq p_1$ such that :

the ball $B(y_{i(1)}, \varepsilon/5)$ contains only one point $a_i, i \in \{1, 2, \ldots, k\}$.

We denote by p_2 the smallest such p .
Let $i'(2) \in \{1, 2, \ldots, k'\}$ be such that $x^2{}_p \in B(y_{i'(2)}, \varepsilon/5)$.

Of course $i'(2) \neq i(1)$, and thus, by recurrence :

$$\forall p \geq p_2   x^2{}_p \in B(y_{i'(2)}, \varepsilon/5) .$$

Thus the sequence $(x^2{}_p)$ converges to $y_{i'(2)}$.

Analogously, we show that $(x^3{}_p)$ either converges or has two accumulation points $y_{i(1)}$ and $y_{i(2)}$ (or $y_{i'(2)}$) ; and so on, until k' convergent sequences are obtained. One notes that $(x^1{}_p)$ and $(x^2{}_p)$ always converge, but $(x^i{}_p)$ $i \geq 3$ does not necessarily converge.

Remarks

Using a "composition" of an algorithm which searches for the number of accumulation points with a U-algorithm, one can extract exactly k subsequences, when (x_n) has k accumulation points. See also [3], p. 166-174.

4 - LIMITATION RESULTS

Two limitation results are presented here : the first one (theorem 6) is devoted to algorithms extracting only one subsequence, the second (theorem 7) is devoted to simultaneous extraction algorithms. These results show that it is not possible to remove hypothesis of our theorems in section 1, 2 and 3 (but perhaps one can replace certain hypothesis by others).

Theorem 6

If card(E) $\geq$ 2 , there is no algorithm A such that, for every sequence $(x_n) \in E^N$ having accumulation points, the transformed sequence (t_n) is a convergent subsequence of (x_n).

Remarks

1)   One can improve the theorem as following :

(a)   by replacing "for every sequence $(x_n) \in E^N$ having accumulation points" by "for every sequence $(x_n) \in E^N$ having one or two accumulation points"

(b)   by replacing "for every sequence $(x_n) \in E^N$ having accumulation points" by "for every sequence $(x_n) \in E^N$ having exactly two accumulation points".

(c)   by replacing "the transformed sequence (t_n) is a convergent subsequence of (x_n)" by "the transformed sequence (t_n) is a sequence which converges to $x \in A(x_n)$".

An easy adaptation of the next proof gives   (a)   and   (c) .   Improvement   (b)   is rather difficult.

2)   Theorem 6 means that no choice of i_0 , $\alpha(n)$, $\beta(n)$ is satisfactory for every sequence (x_n) , but it also means that there is no algorithm (even a more complicated one) which can extract convergent subsequence from every sequence having accumulation points.

Proof

Let $a, b \in E$, $a \neq b$.

Assume that F is an algorithm such that, for every $(x_n) \in E^{\mathbb{N}}$ having accumulation points, the transformed sequence is a convergent subsequence of (x_n).

Let $(x^0_n) = (a, a, a, \ldots)$.

We denote by (t^0_n) the transformed sequence. Necessarily, $t^0_n = a$ for every n .

Set $m(0) = 0$ and let $n(0)$ be the largest index of points x^0_n utilized in the calculation of $t^0_{m(0)}$.

Let (x^1_n) be the sequence :

$$(x^1_n) = (x^0_0, x^0_1, \ldots, x^0_{n(0)}, b, b, \ldots, b, \ldots).$$

We denote by (t^1_n) the transformed sequence. Necessarily, $t^1_{m(0)} = a$ for (x^1_m) has the same $(n(0)+1)$ first points as (x^0_n). From the hypothesis on F , there exists $m(1) > m(0)$ such that $t^1_{m(1)} = b$.

Let $n(1)$ be the largest index of points x^1_n utilized in the calculation of t^1_i , $i \in \{0, 1, \ldots, m(1)\}$.

Let (x^2_n) be the sequence :

$$(x^2_n) = (x^1_0, x^1_1, \ldots, x^1_{n(1)}, a, a, \ldots, a, \ldots);$$

We denote by (t^2_n) the transformed sequence. Necessarily , $t^2_{m(0)} = a$ and $t^2_{m(1)} = b$. There exists $m(2) > m(1)$ such that $t^2_{m(2)} = a$. etc...

We consider the sequence :

$$(x_n) = (x^0_0, x^0_1, \ldots, x^0_{n(0)}, x^1_{n(0)+1}, \ldots, x^1_{n(1)}, x^2_{n(1)+1}, \ldots, x^2_{n(2)}, \ldots)$$

We denote by (t_n) the transformed sequence.

By construction :

$$\left|\begin{array}{l} t_{m(2i)} = a, \\[2mm] t_{m(2i+1)} = b , \end{array}\right.$$

and thus (t_n) is not convergent. This is a contradiction.

Theorem 7

Let $k \in \mathbb{N}$, $k \geq 2$.

Let E be a metric space having at least one accumulation point. There is no algorithm (from $E^{\mathbb{N}}$ into $(E^k)^{\mathbb{N}}$) such that, for every bounded sequence (x_n) with exactly k accumulation points, the algorithm gives k subsequences $(x^1_n), (x^2_n), \ldots, (x^k_n)$ converging to the k accumulation points of (x_n).

Remarks

1) This result shows that, even if the number of accumulation points of (x_n) is known, it is not possible, in general, to extract convergent subsequences for each accumulation point.

2) Theorem 7 does not hold if E has no accumulation points, that is to say, if each point of E is isolated. Indeed, an extraction algorithm is easily constructed from the following characterization :

$$y \in A(x_n) \Longleftrightarrow \{n \in \mathbb{N} \mid x_n = y\} \text{ is infinite.}$$

Proof

We assume $k = 2$ (for other values of k , the proof is similar).

Let (R,C) be an algorithm for sequences such that , for every sequence (x_n) having two accumulation points, it gives two convergent subsequences with different limits.

We shall obtain a contradiction.

Let a be an accumulation point of E and let $(a_0, a_1, a_2, \ldots)$ be a sequence of points different from a , which converges to a .

Let $b \neq a$.

0 - The sequence (x^0_n) is definde by :

$$(x^0_0, x^0_1, x^0_2, x^0_3, \ldots,) = (a, a_0, a, a_0, a, a_0, \ldots)$$

The algorithm (R,C) gives two convergent subsequences (y^0_n) , (z^0_n) with limits a and a_0 .

Let $\theta(0)$ be an integer such that : $y^0_{\theta(0)} \neq z^0_{\theta(0)}$. Until step $\theta(0)$, the algorithm uses only a finite number of points of (x_n) .

Let $\gamma(0)$ be the largest index of points utilized.

For every sequence (x_n) with the same $(\gamma(0)+1)$ first points as (x^0_n) the algorithm gives the same $(\theta(0)+1)$ first answers, hence :

$$\{y_{\theta(0)}, z_{\theta(0)}\} \subset \{a, a_0\} .$$

1 - The sequence (x^1_n) is defined by :

$$x^1_i = x^0_i \text{ for every } i \leq \gamma(0)$$

$$(x^1_{\gamma(0)+1} , x^1_{\gamma(0)+2}, \ldots) = (b, a, a_1, a, a_1, a, a_1, \ldots)$$

The algorithm (R,C) gives two convergent sequences (y^1_n) , (z^1_n) with limits a and a_1 .

There exists an integer $\theta(1) > \theta(0)$ such that :

$$\{y^1_{\theta(1)}, z^1_{\theta(1)}\} \subset \{a, a_1\} .$$

Until step $\theta(1)$, the algorithm use only a finite number of points of (x^1_n) . Let $\gamma(1) > \gamma(0)$ be the largest index of points utilized.

For every sequence (x_n) with the same $(\gamma(1)+1)$ first points as (x^1_n) , the algorithm gives the same $(\theta(1)+1)$ first answers, hence

$$\{y_{\theta(0)}, z_{\theta(0)}\} \subset \{a, a_0\} ,$$

$$\{y_{\theta(1)}, z_{\theta(1)}\} \subset \{a, a_1\} .$$

2 - The sequence (x^2_n) is defined by :

$$x^2_i = x^1_i \text{ for every } i \leq \gamma(1) ,$$

$$(x^2_{\gamma(1)+1}, x^2_{\gamma(1)+2}, \ldots) = (b, a, a_2, a, a_2, a, a_2, \ldots).$$

The algorithm (R,C) gives two convergent sequences $(y^2_n), (z^2_n)$ with limits a, a_2 .

There exists an integer $\theta(2) > \theta(1)$ such that :

$$\{y^2_{\theta(2)}, z^2_{\theta(2)}\} \subset \{a, a_2\}$$

Until step $\theta(2)$, the algorithm use only a finite number of points of (x^2_n) . Let $\gamma(2) > \gamma(1)$ be the largest index of points utilized.

For every sequence (x_n) with the same $(\gamma(2)+1)$ first points as (x^2_n) the algorithm gives the same $(\theta(2)+1)$ first answers, hence :

$$\{y_{\theta(0)}, z_{\theta(0)}\} \subset \{a, a_0\}$$

$$\{y_{\theta(1)}, z_{\theta(1)}\} \subset \{a, a_1\}$$

$$\{y_{\theta(2)}, z_{\theta(2)}\} \subset \{a, a_2\}$$

Etc...

When sequences $(x^0_n),(x^1_n),\ldots$ are constructed, and $\theta(0),\theta(1),\ldots,$ $\gamma(0),\gamma(1),\ldots$ are determined, we set :

$$(x_n) = (x^0_0, x^0_1, \ldots, x^0_{\gamma(0)}, x^1_{\gamma(0)+1}, \ldots, x^1_{\gamma(1)}, x^2_{\gamma(1)+1}, \ldots)$$

This sequence has two accumulation points a and b, hence one of the two sequences given by (R,C) must converge to b.

This is impossible, for :

$$\{y_{\theta(0)}, z_{\theta(0)}\} \subset \{a, a_0\}$$

$$\{y_{\theta(1)}, z_{\theta(1)}\} \subset \{a, a_1\}$$

$$\{y_{\theta(2)}, z_{\theta(2)}\} \subset \{a, a_2\}$$

...

REFERENCES

[1]   BERGE C.
        "Espaces topologiques – Fonctions multivoques"
        Dunod, Paris, 1966.

[2]   BERTIER P. et BOUROCHE J.M.
        "Analyse des données multidimensionnelles"
        Presses Universitaires de France, Paris, 1975.

[3]   DELAHAYE J.P.
        "Quelques problèmes posés par les suites de points non
        convergentes et algorithmes pour traiter de telles suites"
        Thèse de 3ème Cycle, Lille, 1979.

[4]   DELAHAYE J.P.
        "Expériences numériques sur les algorithmes d'extraction
        pour suites non convergentes"
        Publication ANO n° 5, Université des Sciences et Techniques
        de Lille, 1979.

[5]   DELAHAYE J.P.
        "Algorithmes–questions et algorithmes d'extraction pour
        suites non convergentes"
        Bulletin de la Direction des Etudes et Recherches de l'E.D.F.,
        C, n° 1, 1979, pp. 17–34.

[6]   DELAHAYE J.P.
        "Algorithmes pour suites non convergentes"
        Numer. Math., 34, 1980, pp. 333–347.

[7]   DELAHAYE J.P.
        "Théorèmes de points fixes centrés"
        Publication ANO n° 25, Université des Sciences et Techniques
        de Lille, 1980.

[8]   DELAHAYE J.P.
        "Algorithmes pour extraire une sous–suite convergente d'une
        suite non convergente"
        Proceeding of "Conference optimization : theory and
        algorithms"
        Lecture Notes in Pure et Applied Mathematics,
        Marcel Dekker, New York, 1983.

[9]   DELAHAYE J.P.
        "The cluster pont set of a non–convergent sequence"
        Publication ANO n° 68, Université des Sciences et Techniques
        de Lille, 1982.

[10]   DELAHAYE J.P. et DENEL J.
        "Equivalence des continuités des applications multivoques
        dans des espaces topologiques"
        Publication n° 111 du Laboratoire de Calcul de l'Université
        des Sciences et Techniques de Lille, 1978.

[11]   DELAHAYE J.P. et DENEL J.
        "The continuities of point-to-set maps, definitions and
        equivalences"
        Mathematical Programming Study, 10, 1979, pp.8-12.

[12]   DENEL J.
        "Extension of the continuity of point-to-set maps :
        applications to fixed point algorithms"
        Mathematical Programming Study 10, 1979, pp. 48-68.

[13]   EAVES B.C.
        "Non linear programming via Kakutani fixed points"
        Working paper n° 294, Center for Research in Management
        Science, University of California, Berkeley, 1976.

[14]   EAVES B.C.
        "Computing Kakutani Fixed Points"
        SIAM J. Appl. Math. 21, 1971, pp. 236-244.

[15]   EAVES B.C.
        "Homotopies for computation of fixed points"
        Mathematical Programming 3, 1972, pp. 1-22.

[16]   EDELSTEIN M.
        "On fixed and periodic points under constructive mappings"
        Jour. London Math. Soc. 37, 1962, pp. 74-79.

[17]   FIOROT J.C. et HUARD P.
        "Composition and union of general algorithms of optimization"
        Mathematical Programming Study, 10, 1979, pp. 69-85.

[18]   HUARD P.
        "Optimisation dans R^n"
        Cours de D.E.A. polycopié. Université des Sciences et
        Techniques de Lille, 1972.

[19]   HUARD P.
        "Extensions of Zangwill's Theorem"
        Mathematical Programming Study 10, 1979, pp. 98-103.

[20]   KAKUTANI S.
        "A generalization of Brouwer's fixed point theorem"
        Duke Math. J. 8, 1941, pp. 457-459.

[21]   MEYER G.G.L.
"Asymtotic Properties of Sequences Iteratively Generated
by Point-to-set Maps"
Mathematical Programming Study 10, 1979, pp. 115-125.

[22]   MEYER G.G.L. et RAUP R.C.
"On the structure of cluster points sets of iteratively
generated sequences"
Jour. of Optimization Theory and Applications, 28, 1979,
pp. 353-362.

[23]   POLAK E.
"Computational Methods in Optimization : A unified Approach"
Academic Press, New-York, 1971.

[24]   SAIGAL R.
"The fixed point approach to nonlinear programming"
Mathematical Programming Study 10, 1979, pp. 142-157.

[25]   SCARF A.
"The approximation of fixed points of a continuous mapping"
SIAM J. Appl. Math. 15, 1967, pp. 1328-1343.

[26]   SHARKOVSKII A.N.
"Attracting and Attracted Sets"
Soviet Math. 6, 1965, pp. 268-270.

[27]   TODD M.J.
"The computation of fixed points and applications"
Springer-Verlag, Lecture Notes in Economics and
Mathematical Systems, 124, 1976.

[28]   ULAM S. et STEIN P.
"Non-linear Transformation Studies on Electronic Computers"
Rozprawy Matematyczne 39, 1964, pp. 3-66.

[29]   ZANGWILL W.E.
"Nonlinear Programming : A unified approach"
Prentice Hall, Englewood Cliffs, 1969.

Chapter 4

The Partially Ordered Systems
of Accelerable Families

INTRODUCTION

In this chapter, we introduce and study the general aspects of the notions used in the acceleration of convergence.

In section 1, we define the acceleration velocity, and the asymptotic acceleration velocity. This gives the usual notion of acceleration of the convergence ([1],[14]). We recall the notion of degree of acceleration ([11]) and the notion of predicted sequence.

In section 2, we are concerned with families of sequences related to the acceleration. We obtain a few ordered systems of sequences families.

In section 3, we give examples, then in section 4 we study the relations between the ordered systems previously defined.

Section 5 is devoted to the study of the existence of maximal accelerable families, and other related problems.

Notation

$E^{\mathbf{N}}$: the set of infinite sequences of elements of E

$E^{(\mathbf{N})}$: the set of finite sequences of elements of E

$\mathrm{Conv}(E)$: the set of convergent sequences of E

If $(x_n),(y_n),(x^0{}_n),(x^1{}_n),\ldots,(x^i{}_n) \in \mathrm{Conv}(E)$

their limits are respectively denoted by
$x, y, x^0, x^1, \ldots, x^i$:

$\mathrm{Conv}^*(E)$: The set of sequences $(x_n) \in \mathrm{Conv}(E)$ such that :

$\exists\, n_0 \in \mathbf{N}$, $\forall\, n \geq n_0 : x_n \neq x$

$\mathrm{Conv}_0(R^+) = \{(x_n)\,|\,(x_n) \in \mathrm{Conv}(R) \ \text{ and } \ x = 0\}$

$P(E)$: the set of all the subsets of E.

1 - ACCELERATION VELOCITY, ACCELERATION, PREDICTION

This section deals with five notions of acceleration which are used later. In addition to the usual one ([1],[11],[14]), we add the notion of acceleration with the velocity (ε_n) and the notion of acceleration with the asymptotic velocity (ε_n), which permit us to measure the quality of the acceleration provided by a sequence (t_n) relatively to a sequence (x_n).

The two other notions are the notion of degree of acceleration (first introduced in [11]) and the notion of prediction which expresses the exactness of sequence transformations for sequences.

Definitions

E is a metric space with distance d .

Let $(x_n) \in \text{Conv}(E)$, $(\varepsilon_n) \in \text{Conv}_0(R^+)$

We say that the sequence $(t_n) \in \text{Conv}(E)$ converge more rapidly than the sequence (x_n) with the velocity (ε_n) (resp. with the asymptotic velocity (ε_n)) , if and only if (x is the limit of (x_n)) :

$$\forall\ n \in \mathbf{N} : d(t_n,x) \leq \varepsilon_n\ d(x_n,x)$$

(resp. $\dashv\ n_0 \in \mathbf{N}$, $\forall\ n \geq n_0 : d(t_n,x) \leq \varepsilon_n\ d(x_n,x))$.

We say that the sequence $(t_n) \in \text{Conv}(E))$ converges more rapidly than the sequence (x_n), if and only if there exists $(\varepsilon_n) \in \text{Conv}_0(R^+)$ such that (t_n) converges more rapidly than the sequence (x_n) with the velocity (ε_n) [1] . This condition may be written :

$$\forall\ \varepsilon \in R^{+*} ,\ \dashv\ n_0 \in \mathbf{N} ,\ \forall\ n \geq n_0 : d(t_n,x) \leq \varepsilon\ d(x_n,x) .$$

When $(x_n) \in \text{Conv}^*(E)$, this condition may also be written :

$$\lim_{n\to\infty} d(t_n,x)\ /\ d(x_n,x) = 0 .$$

We say that the sequence (t_n) predicts the limit of the sequence $(x_n) \in \text{Conv}(E)$ (or predicts (x_n)) , if and only if :

(t_n) converges more rapidly than the sequence (x_n) with the asymptotic velocity $(\varepsilon_n) = (0,0,\ldots,0,\ldots)$.

This condition may be written :

$$\dashv\ n_0 \in \mathbf{N} ,\ \forall\ n \geq n_0 : t_n = x .$$

----

[1]   If we write "with the asymptotic velocity" we obtain an equilavent definition.

For every $\lambda \in \mathbf{R}$, $\lambda \geq 1$ we say that the sequence $(t_n) \in \text{Conv}(E)$
converges more rapidly than the sequence (x_n) with degree λ , if
and only if :

$$\forall \varepsilon \in \mathbf{R}^{+*} , \; \exists n_0 \in \mathbf{N} , \; \forall n \geq n_0 : d(t_n,x) \leq \varepsilon(d(x_n,x))^\lambda .$$

Hence acceleration with degree 1 is the same as acceleration.

Examples

The sequence $(1/2^n)$ converges more rapidly than $(1/(n+1))$ with the
velocity $(\varepsilon_n) = ((n+1)/(2^n))$ and with the asymptotic velocity
$((2/3)^n)$.

The sequence $(1/2,3/4,1,1,1,\ldots)$ predicts the limit of $(1-1/(2^{n+1}))$.

The sequence $(1/5^n)$ converges more rapidly than the sequence $(1/2^n)$
with degree 2.

2 - TRANSFORMATIONS FOR CONVERGENCE ACCELERATION; ACCELERABLE FAMILIES

Each of the definitions given in section 1 may be adjusted (using the
notion of normal transformation) to sequence families. This allows us
to speak of accelerable families with the velocity (ε_n) , of predictible families, etc... The ordered (by the relation of inclusion)
systems of families defined in this way will be studied later.

Definition

Let T be a transformation from $E^{\mathbf{N}}$ into $E^{\mathbf{N}}$ defined on
$S \subset \text{Conv}(E)$ $(T \in \text{Trans}(E,E,))$.

We say that T is _regular_ on **S** , if and only if, for every sequence
$(x_n) \in$ **S**, the transformed sequence of (x_n) by T is convergent
with the same limit as (x_n) .

Let $(\varepsilon_n) \in \text{Conv}_0(\mathbf{R}^+)$.

We say that T accelerates the convergence of **S** (resp. accelerates
the convergence of **S** with the velocity (ε_n) , resp. accelerates
the convergence of **S** with the asymptotic velocity (ε_n); resp.
accelerates the convergence of **S** with degree λ , resp. predicts
S), if, for every sequence $(x_n) \in$ **S** , the transformed sequence of
(x_n) by T converges more rapidly than (x_n) (resp. converges more
rapidly than (x_n) with the velocity (ε_n) ; resp. converges more
rapidly than (x_n) with the asymptotic velocity (ε_n) ; resp.
converges more rapidly than (x_n) with the degree λ ; resp. predicts
the limit of the sequence (x_n)).

If T is a transformation which predicts **S** , we also say that T
is _exact_ on **S.**

Remark

By definition, if T accelerates **S,** then T is regular on **S.**

We say that the family **S** ⊂ Conv(E) is _accelerable_ (resp. is _acce-_
lerable with the velocity (ε_n); resp _is accelerable with the_ _asymp-_
totic velocity (ε_n) , resp. _is accelerable with degre_ λ ; resp. _is_
predictible) if there exists a normal transformation which accelerates the convergence of **S** (resp. accelerates the convergence of **S**
with the velocity (ε_n) ; resp. accelerates the convergence of **S**
with degre λ ; resp. predicts **S**).

The set of accelerable families of Conv(E) is denoted by :

$$\mathbf{A}(E)\ ^{(1)}\ .$$

The set of accelerable families with the velocity (ε_n) is denoted by

$$\mathbf{A}_{(\varepsilon_n)}(E)\ .$$

The set of accelerable families with the asymptotic velocity (ε_n) is
denoted by :

$$\mathbf{A'}_{(\varepsilon_n)}(E)\ .$$

The set of accelerable families with degree λ is denoted by :

$$\mathbf{A}_\lambda(E)\ .$$

The set of predictible families is denoted by :

$$\mathbf{D}(E)\ .$$

Remarks

1)   We have to note that, in the definition, only normal   (i.e.
0-normal) transformations are involved. This restriction seems a very
natural one to us, for when concerned with the acceleration of
convergence, what is of importance is the progress given by t_n
compared with the last term used in the calculation of t_n. Even-
tually, with the help of normalization, we can always suppose that
this last term is x_n.   The Aitken's Δ^2-Process is a good illus-
tration of the question (see the remark of [5], p. 324).

----

(1)   don't confuse **A**(E) with the set of accumulation points of
       (x_n) : $\mathbf{A}(x_n)$.

2) To try to predict a family $S \subset Conv(E)$ is equivalent to looking for an algorithm question satisfactory for the question " what is the limit of (x_n) ?" . The theorem on p. ... allows us to state the equivalence between :

    * "There exists $A \subset {}^{\tau}Alg(E,E)$ which predicts S"

and
    * "There exists $A \subset {}^{\tau}Norm(E,E)$ which predicts S".

This situation is a very special one and as we shall see later does not hold with the other notions.

3) The identity transformation is regular for every sequence $(x_n) \in Conv(E)$. This is the reason why we have not defined the notion of regular family.

3 - EXAMPLES OF ACCELERABLE FAMILIES OF SEQUENCES

a) Let K be R or C.

Let $\lambda \in K$, $\lambda \neq 0$, $\lambda \neq 1$. We define :

    $Lin_\lambda(K) = \{(x_n) \in Conv^*(K) | \lim_{n \to \infty}(x_{n+1}-x)/(x_n-x) = \lambda\}$;

    $Lin^{\Delta}_\lambda(K) = \{(x_n) \in Conv^*(K) | \lim_{n \to \infty}(x_{n+2}-x_{n+1})/(x_{n+1}-x_n) = \lambda\}$;

(If $|\lambda| < 1$ then $Lin_\lambda(K) = Lin^{\Delta}_\lambda(K)$;

if $\lambda = -1$ then $Lin_\lambda(K) \subset Lin^{\Delta}_\lambda(K)$;   (see chapter 6 and $[3],[4],[14]$).

    $Lin(K) = \cup \{Lin_\lambda(K) | 0 < |\lambda| \leq 1, \lambda \neq 1\}$;

Every $Lin_\lambda(K)$ with $0 < |\lambda| \leq 1$, $\lambda \neq 1$ is accelerable; $Lin(K)$ is even accelerable by the Aitken Δ^2-Process.

There by we can write :

$$Lin(K) \in A(K) .$$

b) Let $(\varepsilon_n) \in Conv_0(R^+)$. We define :

    $S = \{(x_n) \in Conv(R) \ | \ \forall n \in N^* : x_{n-1}-x \neq 0$
    and
$$\left| \frac{x_{n-1}-x}{x_n-x} +1 \right| \leq 2 \varepsilon_n \}$$

The normal transformation defined by :

$$t_n = (x_n + x_{n-1})/2$$

accelerates **S** with the velocity (ε_n). Indeed :

$$\left| \frac{t_n - x}{x_n - x} \right| = \left| \frac{(x_n - x) + (x_{n-1} - x)}{2(x_n - x)} \right| \leq \varepsilon_n$$

With our notation, we write :

$$S \in A_{(\varepsilon_n)}(R) .$$

c) Let $(\alpha_n) \in Conv_0(R^+)$ be such that $\alpha_n \neq \alpha_m$ for $n \neq m$. We define :

$$S = \{(x_n) \in Conv(R) \mid \exists N \in \mathbf{N} , \exists(\lambda_0, \ldots, \lambda_N) \in R^{N+1}$$

$$\forall n \in \mathbf{N} : x_n = \sum_{i=0}^{N} \lambda_i \, \alpha^i_n\}$$

The Richardson process ([1],[2]) is an exact transformation for **S**. With our notation, we write :

$$S \in D(R) ,$$

which is also equivalent to :

$$S \in A'_{(\varepsilon_n)}(R) ,$$

where $(\varepsilon_n) = (0, 0, \ldots, 0, \ldots)$.

d) Let $S = \{(x^i_n) \mid i \in \mathbf{N}\} \subset Conv(E)$ be an arbitrary denumerable family of convergent sequences.

Define the normal algorithm (f_n) as following :

* $f_i(x_0, x_1, \ldots, x_i) = x^j$ where j is the least integer $k \leq i$ such that $(x_0, x_1, \ldots, x_i) = (x^k_0, x^k_1, \ldots, x^k_i)$, if such an integer exists,

* $f_i(x_0, x_1, \ldots, x_i) = x_i$ otherwise.

We easily obtain that this algorithm is exact on **S** , which is denoted by :

$$S \in D(E) .$$

One can remark that the algorithm (f_n) is regular on Conv(E) , and that it is more an "identification" algorithm than an authentic acceleration algorithm. Indeed, (f_n) recognizes what sequences we are giving to it, then, after identification, proposes as answer the limit of the recognized sequence.

e) Examples of accelerable families with degree $\lambda \neq 1$ are given in [11].

4 - RELATIONSHIPS BETWEEN THE ORDERED SYSTEMS OF ACCELERABLES FAMILIES

Is it possible to reduce some of our definition to others ?

For example :

* Is every accelerable family an accelerable family with a certain velocity ? (resp. with a certain asymptotic velocity)

* Is every accelerable family with an asymptotic velocity (ε_n) an accelerable family with a velocity (ε'_n) ?

What are the relationships between accelerable families with a given velocity and predictible families.

In this section, we try to answer such questions. Simultaneously, we give examples showing the interest of the five introduced notions, and we establish some characterizations.

The Theorem 1 follows immediately from the definitions :

Theorem 1

(i) $\forall\ (\varepsilon_n) \in \mathrm{Conv}_0(\mathbf{R}^+)$, $\forall\ \lambda \in [1,\infty[$:

$$\mathbf{P}(\mathrm{Conv}(\mathbf{R}))$$

$$\cup$$

$$\mathbf{A}(\mathbf{R}) \supset \mathbf{A}_\lambda(\mathbf{R})$$

$$\cup$$

$$\mathbf{D}(\mathbf{R}) \subset \mathbf{A}'_{(\varepsilon_n)}(\mathbf{R}) \supset \mathbf{A}_{(\varepsilon_n)}(\mathbf{R})$$

(ii) $\forall\ (\varepsilon_n),(\eta_n) \in \mathrm{Conv}_0(\mathbf{R}^+)$:

$$(\forall\ n \in \mathbf{N} : \varepsilon_n \geq \eta_n) \Longrightarrow \left|\ \begin{array}{l} \mathbf{A}_{(\varepsilon_n)}(\mathbf{R}) \supset \mathbf{A}_{(\eta_n)}(\mathbf{R}) \\[2ex] \mathbf{A}'_{(\varepsilon_n)}(\mathbf{R}) \supset \mathbf{A}'_{(\eta_n)}(\mathbf{R}) \end{array}\right.$$

(iii) $\forall \lambda, \lambda' \in [1, \infty[$; $\lambda < \lambda'$:

$$A_\lambda(R) \supset A_{\lambda'}(R)$$

(iv) $\forall S \in A(R)$: $S' \subset S \Rightarrow S' \in A(R)$

(and the same holds with :

$A'_{(\varepsilon_n)}(R)$, $A_{(\varepsilon_n)}(R)$, $A_\lambda(R)$ and $D(R)$) .

In the remainder of this section, we will assume that $E = R$ and give complementary results concerning extensions.

Theorem 2

There exist non-accelerable families of convergent real sequences :

$$A(R) \neq P(\mathrm{Conv}(R)) .$$

The proof of theorem 2 is given in chapter 5 (and $[9], [10]$).

Remark

Theorem 2 holds for every metric space E with at least one accumulation point.

Let **S** be an arbitrary accelerable family. Is there an acceleration velocity naturaly associated with **S** ? Generally, the answer is NO :

Theorem 3

$$\cup \{ A_{(\varepsilon_n)}(R) \,|\, (\varepsilon_n) \in \mathrm{Conv}_0(R^+) \} \subset_{\neq} A(R)$$

$$\cup \{ A'_{(\varepsilon_n)}(R) \,|\, (\varepsilon_n) \in \mathrm{Conv}_0(R^+) \} \subset_{\neq} A(R)$$

Remark

A family belonging to **A(R)** but not to any $A_{(\varepsilon_n)}(R)$ (Lemma 1 gives gives an example) is an accelerable family which has badly accelerated sequences for every normal algorithm.

Proof

This follows immediatly from lemma 1.

Lemma 1

$\forall \ \lambda \in \mathbf{R} \ , \ 0 < \lambda < 1 \ , \ \forall \ (\varepsilon_n) \in \text{Conv}_0(\mathbf{R}^+) : \text{Lin}_\lambda(\mathbf{R}) \not\subset A'_{(\varepsilon_n)}(\mathbf{R})$

Proof of Lemma 1

Let $(\varepsilon_n) \in \text{Conv}_0(\mathbf{R}^+)$ and $A = (f_n)$ be a normal transformation such that, for every sequence $(x_n) \in \text{Lin}_\lambda(\mathbf{R})$, the transformed sequence (t_n) accelerates (x_n) with the asymptotic velocity (ε_n).

0°) We define $(x^0{}_n) \in \text{Lin}_\lambda(\mathbf{R})$ by $x^0{}_0, x^0{}_1$, such that $x^0{}_0 < x^0{}_1$ and by the formula :

$$(x^0{}_{n+2} - x^0{}_{n+1})/(x^0{}_{n+1} - x^0{}_n) = \lambda$$

This sequence converges to $x^0 = x_0 + (x_1 - x_0)/(1 - \lambda)$.

From the hypothesis, there exists $n_0 \in \mathbf{N}$ such that :

$$|t^0{}_{n_0} - x^0| \le \varepsilon_{n_0}|x^0{}_{n_0} - x^0| \quad \text{and} \quad \varepsilon_{n_0} < 1/2^1$$

1°) We define $(x^1{}_n) \in \text{Lin}_\lambda(\mathbf{R})$ by :

$$x^1{}_n = x^0{}_0 \quad \text{for} \quad n \le n_0$$

$$x^1{}_n = x^0{}_n + 4 \ (x^0 - x^0{}_{n_0})/2^1$$

This sequence converges to $x^1 = x^0 + 4(x^0 - x^0{}_{n_0})/2^1$

It is easy to show that, for all $x \ge x^1$:

$$|t^0{}_{n_0} - x|/ \ |x^0{}_{n_0} - x| \ge 1/2 > \varepsilon_n \quad .$$

From the hypothesis, there exists $n_1 > n_0$ such that :

$$|t^1{}_{n_1} - x^1| \le \varepsilon_{n_1}|x^1{}_{n_1} - x^1| \quad \text{and} \quad \varepsilon_{n_1} < 1/2^2$$

2°) We define $(x^2{}_n) \in \text{Lin}_\lambda(\mathbf{R})$ by :

$$x^2{}_n = x^1{}_n \quad \text{for} \quad n \le n_1$$

$$x^2{}_n = x^1{}_n + 4 \ (x^1 - x^1{}_{n_1})/2^2$$

This sequence converges to

$$x^2 = x^1 + 4 \ (x^1 - x^1{}_{n_1})/2^2 \quad .$$

It is easy to show that, for all $x \geq x^2$:

$$|t^1{}_{n_1} - x| \ / \ |x^1{}_{n_1} - x| \geq 1/2^2 > \varepsilon_{n_1} \ .$$

From the hypothesis, there exists $n_2 > n_1$ such that :

$$|t^2{}_{n_2} - x^2| \leq \varepsilon_{n_2}|x^2{}_{n_2} - x^2| \quad \text{and} \quad \varepsilon_{n_2} < 1/2^3$$

...

i+1°) We define $(x^{i+1}{}_n) \in \text{Lin}_\lambda(R)$ by :

$$x^{i+1}{}_n = x^i{}_n \quad \text{for} \quad n \leq n_i$$

$$x^{i+1}{}_n = x^i{}_n + 4 \ (x^i - x^i{}_{n_i})/2^{i+1}$$

This sequence converges to

$$x^{i+1} = x^i + 4 \ (x^i - x^i{}_{n_i})/2^{i+1}$$

It is easy to show that, for all $x \geq x^{i+1}$:

$$|t^i{}_{n_i} - x| \ / \ |x^i{}_{n_i} - x| \geq 1/2^{i+1} > \varepsilon_{n_i}$$

From the hypothesis, there exists $n_{i+1} > n_i$ such that :

$$|t^{i+1}{}_{n_{i+1}} - x^{i+1}| \leq \varepsilon_{n_{i+1}} |x^{i+1}{}_{n_{i+1}} - x^{i+1}| \quad \text{and} \quad \varepsilon_{n_{i+1}} \leq 1/2^{i+2} \ .$$

...

The sequence $(x_n) = (x^0{}_0, x^0{}_1, \ldots, x^0{}_{n_0}, x^1{}_{n_0+1}, \ldots, x^1{}_{n_1}, x^2{}_{n_1+1}, \ldots)$
is transformed into the sequence :

$$(t_n) = (t^0{}_0, t^0{}_1, \ldots, t^0{}_{n_0}, t^1{}_{n_0+1}, \ldots, t^1{}_{n_1}, t^2{}_{n_1+1}, \ldots)$$

The limit x of the sequence (x_n) satisfies :

$$x > x^i \quad \text{for all} \quad i \in \mathbf{N} \ ,$$

and, consequently :

$$|t_{n_i} - x| \ / \ | x_{n_i} - x| > \varepsilon_{n_i} \quad \text{for all} \quad i \in \mathbf{N}$$

Since, we have $(x_n) \in \text{Lin}_\lambda(R)$, we obtain a contradiction.

Remark

Theorem 3 is still true when R is replaced by $E \subset R^n$, $E \neq \emptyset$.

Let S be an accelerable family with the asymptotic velocity (ε_n) . Is it possible to find a sequence (η_n) (slower than (ε_n)) such that S is accelerable with the velocity (η_n) ?

Generally, we still have a negative answer.

Theorem 4

For every $(\varepsilon_n) \in \text{Conv}_0(R)$:

$$\cup \{ A_{(\eta_n)}(R) \mid (\eta_n) \in \text{Conv}_0(R^+) , (\varepsilon_n) \leq (\eta_n) \} \subsetneq A'_{(\varepsilon_n)}(R)$$

Proof

Let $(\varepsilon_n) \in \text{Conv}_0(R^+)$. We define :

$$T_{(\varepsilon_n)} = \{ (x_n) \in \text{Conv}(R) \mid \exists\, n_0 \in N , \forall\, n \geq n_0 \,|x_n + \frac{1}{n+1} - x|$$

$$\leq \varepsilon_n \,|x_n - x| \}.$$

The transformation T defined by $t_n = x_n + \dfrac{1}{n+1}$ accelerates $T_{(\varepsilon_n)}$ with the asymptotic velocity (ε_n), which is denoted by :

$$T_{(\varepsilon_n)} \in A'_{(\varepsilon_n)}(R) .$$

The following lemma will permit us to show that $T_{(\varepsilon_n)}$ does not belong to any $A_{(\eta_n)}(R)$.

Lemma 2 (necessary condition for a family to belong to $A_{(\eta_n)}(R)$)

If $S \in A_{(\eta_n)}(R)$, $k \in N$, (x_n) , $(x^1_n) \in S$ then :

$$(\forall\, n \leq k : x_n = x^1_n) \Rightarrow (|x - x^1| \leq \eta_k \,(|x_k - x| + |x^1_k - x^1|))$$

Proof of the lemma 2

Let T be a normal transformation accelerating $A_{(\eta_n)}(R)$ with the velocity (η_n). By hypothesis :

$$
\left|
\begin{array}{l}
|t_k - x| \le \eta_k \, |x_k - x| \\[2mm]
|t^1_k - x^1| \le \eta_k \, |x^1_k - x^1| \\[2mm]
t_k = t^1_k
\end{array}
\right.
$$

Hence :

$$|x-x^1| \le |x-t_k| + |t^1_k-x^1| \le \eta_k \, (|x_k-x| + |x^1_k-x^1|) .$$

Now let $(\eta_n) \in \text{Conv}_0(R^+)$. We will show that :

$$T_{(\varepsilon_n)} \notin A_{(\eta_n)}(R).$$

There exists n_0 such that $\eta_{n_0} < 1$. For $a,b \in R$, $a \ne b$ we define the sequences (x_n) and (x^1_n) by :

$$x_n = a = x^1_n \quad \text{if} \quad n \le n_0$$

$$x_n = a - \frac{1}{n+1} \quad , \quad x^1_n = b - \frac{1}{n+1} \quad \text{if} \quad n > n_0$$

We apply lemma 2 with $k = n_0$, which yields :

$$|a-b| \le \eta_{n_0} \, |a-b|,$$

which is impossible, since we have $\eta_{n_0} < 1$.

Remark

Theorem 4 is still true when R is replaced be $E \subset \overset{\circ}{R^n}$, $E \ne \emptyset$.

Depending on the properties of the sequence (ε_n), an accelerable family of sequences with the velocity (ε_n) is predictible or not :

Theorem 5

Let $(\varepsilon_n) \in \text{Conv}_0(R^+)$:

(i) If there exists an infinite set of integers such that $\varepsilon_n = 0$,
then :

$$A'_{(\varepsilon_n)}(R) = D(R) \ ,$$

(ii) if not :

$$A'_{(\varepsilon_n)}(R) \underset{\neq}{\supset} D(R) \ .$$

Proof

(i) Suppose $\varepsilon_n = 0$ for $n \in \{n_0, n_1, \ldots, n_i, \ldots\}$ with
$n_0 < n_1 < \ldots < n_i < \ldots$.

Let $S \in A'_{(\varepsilon_n)}(R)$ be given . We shall show that $S \in D(R)$.

There exists a normal algorithm $T = (f_n)$ which accelerates the
convergence of S with the asymptotic velocity (ε_n).

We define the normal algorithm $T' = (f_n)$ by :

$$f'_j = f_j \quad \text{if} \quad j < n_0$$

$$f'_j = f_{n_i} \quad \text{if} \quad n_i \leq j < n_{i+1}$$

Now we prove that T' predicts S . For every $(x_n) \in S$ the
hypothesis gives us that :

$$t_{n_i} = x \quad \text{for} \quad i \geq i_0$$

Thus (using the definition of T') :

$$t'_j = x \quad \text{for} \quad j \geq n_{i_0}$$

(ii) Let $(\varepsilon_n) \in \text{Conv}_0(R^+)$ be a sequence such that :
$\exists N \in \mathbf{N}$, $\forall n \geq N : \varepsilon_n \neq 0$.

Without loss of generality, we can assume that :

$$\forall n \in \mathbf{N} : 0 < \varepsilon_n < 1 \ ;$$

and that (ε_n) is monotone decreasing.

Indeed : if (ε_n) and (ε'_n) are different in only a finite
number of terms then $A'_{(\varepsilon_n)}(R) = A'_{(\varepsilon'_n)}(R)$

If $(\varepsilon'_n) \leq (\varepsilon_n)$, then : $A'_{(\varepsilon'_n)}(R) \subset A'_{(\varepsilon_n)}(R)$

We consider the family :

$$\mathbf{V}_{(\varepsilon_n)} = \{(x_n) \in \mathbf{Conv(R)} \,|\, \forall\, n \in \mathbf{R} : x_n < x$$

and

$$\left| x_n + \frac{1}{n+1} - x \right| < \varepsilon_n |x_n - x| \}$$

It is easy to show that, for every sequence $(x_n) \in \mathbf{Conv(R)}$:

$$(x_n) \in \mathbf{V}_{(\varepsilon)} \iff \left| \begin{array}{l} \forall\, n \in \mathbf{N} : \\[2mm] x_n + \dfrac{1}{(n+1)(1+\varepsilon_n)} < x < x_n + \dfrac{1}{(n+1)(1-\varepsilon_n)} \end{array} \right.$$

The transformation defined by :

$$t_n = x_n + \frac{1}{n+1}$$

shows that : $\mathbf{V}_{(\varepsilon_n)} \in \mathbf{A}_{(\varepsilon_n)}(\mathbf{R})$ and consequently that :

$$\mathbf{V}_{(\varepsilon_n)} \in \mathbf{A'}_{(\varepsilon_n)}(\mathbf{R}) .$$

Now we prove that $\mathbf{V}_{(\varepsilon_n)} \notin \mathbf{D(R)}$. Let $T = (f_n)$ be a normal algorithm which predicts $\mathbf{V}_{(\varepsilon_n)}$.

We define $(x^0{}_n) \in \mathbf{V}_{(\varepsilon_n)}$ by : $x^0{}_n = x^0 - \dfrac{1}{n+1}$

There exists $n_0 \in \mathbf{N}$ such that : $\forall\, n \geq n_0$, $t^0{}_n = x^0$.

We consider the open interval.

$$]\alpha_0, \beta_0[= \bigcap_{n \leq n_0} \left] x^0{}_n + \frac{1}{(n+1)(1+\varepsilon_n)} , x^0{}_n + \frac{1}{(n+1)(1-\varepsilon_n)} \right[$$

Since $x_0 \in]\alpha_0, \beta_0[$, this interval is non-empty.

We choose $x^1 \in](2\,\alpha_0 + \beta_0)/3 , (\alpha_0 + 2\,\beta_0)/3[$, $x^1 \neq x^0$.

We define $(x^1{}_n) \in V_{(\varepsilon_n)}$ by

$$x^1{}_n = x^0{}_n \quad \text{if} \quad n \leq n_0$$
$$x^1{}_n = x^1 - \frac{1}{n+1} \quad \text{if} \quad n > n_0 .$$

There exists $n_1 \in \mathbf{N}$ such that : $\forall\, n \geq n_1$, $t^1{}_n = x^1$.

We consider the open interval :

$$]\alpha_1,\beta_1[\; = \; \underset{n \leq n_1}{\cap} \; \left] x^1{}_n + \frac{1}{(n+1)(1+\varepsilon_n)} \; , \; x^1{}_n + \frac{1}{(n+1)(1-\varepsilon_n)} \right[$$

Since $x_1 \in \,]\alpha_1,\beta_1[$, this interval is non-empty.

We choose :

$$x^2 \in \,](2\,\alpha_1+\beta_1)/3 \; , \; (\alpha_1+2\,\beta_1)/3[\; , \; x^2 \neq x^1 , \; x^2 \neq x^0 ,$$

... ...

The sequence $(x_n) = (x^0{}_0, x^0{}_1, \ldots, x^0{}_{n_0}, x^1{}_{n_0+1}, \ldots, x^1{}_{x_n}, x^2{}_{n_1}+1)$
is transformed into the sequence :

$$t_{(n)} = (t^0{}_0, t^0{}_1, \ldots, t^0{}_{n_0}, t^1{}_{n_0+1}, \ldots, t^1{}_{n_1}, t^2{}_{n_1}+1, \ldots)$$

This sequence (x_n) converges to x , which belongs to :

$$\underset{n \in \mathbf{N}}{\cap} \; \left] x_n + \frac{1}{(n+1)(1+\varepsilon_n)} \; , \; x_n + \frac{1}{(n+1)(1-\varepsilon_n)} \right[.$$

Thus, we have $(x_n) \in V_{(\varepsilon_n)}$.

By definition, (t_n) is not ultimately stationary, consequently T does not predict (x_n) .

Similarly to $\mathbf{A'}_{(\varepsilon_n)}(R)$ in theorem 5 , we now see that, except in special cases, $\mathbf{A}_{(\varepsilon_n)}(R)$ is different from $\mathbf{D}(R)$.

Theorem 6

Let $(\varepsilon_n) \in \text{Conv}_0(R^+)$.

(i)   If there is an integer N such that $\varepsilon_N = 0$, then :

$$A_{(\varepsilon_n)}(R) \underset{\neq}{\subset} D(R)$$

(ii) If not :

$$A_{(\varepsilon_n)}(R) \cap cD(R) \neq \emptyset \quad \text{and} \quad D(R) \cap c\, A_{(\varepsilon_n)}(R) \neq \emptyset .$$

(cE denotes the complementary of E).

Proof

(i)   Let N be an integer such that $\varepsilon_n = 0$.   Let $S \in A_{(\varepsilon_n)}(R)$ be given. We shall show that $S \in D(R)$.   From the hypothesis, there exists a normal algorithm which accelerates the convergence of S with the velocity (ε_n).

We define the normal algorithm $T' = (f'_n)$ by :

$$f'_j = f_j \quad \text{if} \quad j < N$$

$$f'_j = f_N \quad \text{if} \quad j \geq N$$

It is clear that T' predicts S.

(ii) Let $(\varepsilon_n) \in \text{Conv}_0(R^+)$ be a sequence such that

$$\forall\, n \in N : \varepsilon_n \neq 0.$$

Part (ii) of Theorem 4 shows that :

$$A_{(\varepsilon_n)}(R) \cap cD(R) \neq \emptyset.$$

With lemma 2 (in the proof of theorem 3), we easily obtain :

$$D(R) \cap cA_{(\varepsilon_n)}(R) \neq \emptyset .$$

Remarks

1°) Theorems 5 and 6 are still true when R is replaced by $E \subset R^n$ with $E \neq \emptyset$.

2°) The notion of acceleration with degree λ is not reducible to others notions. Indeed, we shall see certain families of sequences which are accelerable with degree 1 and not with degree $1+\varepsilon$ ($\varepsilon > 0$).

5 - MAXIMAL ACCELERABLE FAMILIES

A very natural question is : "do there exist acceleration algorithms which are impossible to improve ?" that is to say such that : "every family of sequences strictly containing the family of sequences they accelerate is not accelerable ?" [12] . This is this problem of maximal families, with which we concern ourselves here.

We shall see that most of the time, there are no maximal accelerable families, or equivalently that almost all acceleration methods may be transformed into others which accelerate strictly large family of sequences.

Theorem 7

$A(R)$ has no maximal families.

$$\forall \ S \in A(R) \ , \ \exists \ S' \in A(R) : S' \supset S \ \text{ and } \ S \neq S'.$$

Proof 1 [13]

Let $S \in A(R)$. Let $F = (f_n)$ be a normal algorithm which accelerates S.

We consider a sequence $(y_n) \notin S$, $(y_n) \in Conv(R)$ (Such a sequence (y_n) exists since $Conv(R)$ is not accelerable).

We define the following normal algorithm $G = (g_n)$ by :

$$g_n(x_0,x_1,\ldots,x_n) = y \quad \text{if} \quad (x_0,x_1,\ldots,x_n) = (y_0,y_1,\ldots,y_n)$$

$$g_n(x_0,x_1,\ldots,x_n) = f_n(x_0,x_1,\ldots,x_n) \quad \text{if not.}$$

Now we show that G accelerates $S \cup \{(y_n)\}$.

If $(x_n) \in S$, there exists n_0 such that, for all $n \geq n_0$:

$$(x_0,x_1,\ldots,x_n) \neq (y_0,y_1,\ldots,y_n) \ ,$$

and consequently, for $n \geq n_0$, F and G give the same transformed point. Thus G accelerates (x_n).

Since G always gives the limit of (y_n) as the transformed point for (y_n) , G is exact and thus G accelerates (y_n).

This shows that S is not maximal.

Proof 2

Let $S \in A(R)$. Let $F = (f_n)$ be a normal algorithm which accelerates S.

We consider a sequence $(y_n) \notin S$, $(y_n) \in Conv(R)$.

We define the following normal algorithm $G = (g_n)$ by :

$$g_n(x_0, x_1, \ldots, x_n) = y \quad if \quad |y-x_n| \leq |y-y_n|$$

$$g_n(x_0, x_1, \ldots, x_n) = f_n(x_0, x_1, \ldots, x_n) \quad if\ not.$$

Now we show that G accelerates $S \cup \{(y_n)\}$.

Let $(x_n) \in S$ be given. If $x \neq y$, then (x_n) is accelerated by G. If $x = y$, then the sequence (t'_n) given by G for (x_n) is such that : $\forall n \in N$, $t'_n = y$ or $t_n = f_n(x_0, x_1, \ldots, x_n)$. Consequently, G accelerates (x_n).

With (y_n) , G is always exact and thus G accelerates (y_n).

Remark

Both "enlargement processes" used in proofs 1 and 2 improve the domain of efficiency of F , but damage the acceleration for certain sequences of S.

The first process damages the acceleration of sequences which have the same beginning as (y_n).

The second process damages the acceleration of sequences which have terms close to y .

Generally, the second process adds more than one sequence to S. In fact, it adds a non-denumerable set of sequences. Indeed, G accelerates all sequences in S and all convergent sequences (x_n) such that $x = y$ and $\exists n_0 \in N$, $\forall n \geq n_0$: $|x_n-x| \leq |y_n-y|$.

The Θ-procedure ($[12]$) is also an "enlargement process", but is of no use here, for it is not certain that it strictly enlarges all S.

Corollary $[13]$

There are no minimal non-accelerable families :

$$\forall S \notin A(R) \ \exists \ S' \notin A(R) : S' \subset S \quad and \quad S' \neq S.$$

Proof

Let S be a non-accelerable minimal family :

$$\forall \ S' \subset S \ , S' \neq S \Rightarrow S' \in A(R)$$

For every $(y_n) \in S$, $S' = S - \{(y_n)\}$ is an accelerable family. Hence, using an enlargement process, we obtain that $S' \cup \{(y_n)\} = S$ is accelerable, which contradicts the definition of S.

Theorem 8

Let $(\varepsilon_n) \in Conv_0(R)$.

$A'_{(\varepsilon_n)}(R)$ does not have maximal families :

$$\forall \ S \in A'_{(\varepsilon_n)}(R) \ , \ \nexists \ S' \in A'_{(\varepsilon_n)}(R) : S' \supset S \text{ and } S' \neq S.$$

Theorem 9

$D(R)$ does not have maximal families :

$$\forall \ S \in D(R) \ , \ \nexists \ S' \in D(R) : S' \supset S \text{ and } S' \neq S.$$

Theorem 10

Let $\lambda \in [1, +\infty[$

$A_\lambda(R)$ does not have maximal families :

$$\forall \ S \in A_\lambda(R) \ , \ \nexists \ S' \in A_\lambda(R) : S' \supset S \text{ and } S' \neq S.$$

Theorems 8, 9, 10 are established in a similar fashion to theorem 7, and analogous corollaries may be obtained.

Remark

Theorems 7, 8, 9 and 10 are still true when R is replaced by $\overset{\circ}{E} \subset R^n$, $E \neq \emptyset$.

We have not given a theorem for $A_{(\varepsilon_n)}(R)$. In fact, for certain sequences $(\varepsilon_n) \in Conv_0(R)$, $A_{(\varepsilon_n)}(R)$ has a maximal family.

We give an example :

$$S = \{(x_n) \in Conv(R) \mid x_0 = x\}$$

$$\varepsilon_n = 0 \quad \text{for} \quad n \in \mathbf{N}.$$

The normal algorithm defined by $t_n = x_0$ accelerates S and hence $S \in A_{(\varepsilon_n)}(R)$. We shall show that S is maximal. Suppose that there

exist $S' \supset S$, $S' \neq S$ and $F = (f_n)$ a normal algorithm which accelerates S' with the velocity (ε_n) .

Let $(y_n) \in S'$, $(y_n) \in S$. We have $y_0 \neq y$, hence $f_0(y_0) = y$ but $(y_0, y_0, \ldots, y_0, \ldots) \in S$ hence $f_0(y_0) = y_0$ which is a contradiction.

The maximal family just considered satisfies $\bar{S} = R^{(N)}$, which is general :

Proposition

Let $(\varepsilon_n) \in \text{Conv}_0(R)$. If S is a maximal family in $A_{(\varepsilon_n)}(R)$,

then $\bar{S} = R^{(N)}$.

We recall that, by definition :

$$\bar{S} = \{s \in E^{(N)} | \; \overline{\exists} \; (x_n) \in S ; \; \overline{\exists} \; n_0 \in N : s = (x_0, x_1, \ldots, x_{n_0})\}$$

Proof

Suppose that $\bar{S} \neq R^{(N)}$.

Let $(a_0, a_1, \ldots, a_n)$ be a sequence not belonging to $\bar{S}$ with minimal length.

Since $(a_0, a_1, \ldots, a_{n-1}) \in \bar{S}$, there exists :

$$(a_0, a_1, \ldots, a_{n-1}, x_n, x_{n+1}, \ldots) \in S .$$

Since the sequence

$$(a_0, a_1, \ldots, a_{n-1}, a_n, x_{n+1}, x_{n+2}, \ldots)$$

begins with $(a_0, a_1, \ldots, a_n)$, it does not belong to S.

Let $F = (f_n)$ be a normal algorithm which accelerates S with the velocity (ε_n).

We define $G = (g_n) \in \mathbf{Norm}(\mathbf{R}, \mathbf{R})$ by :

$g_0 = f_0, \ldots, g_{n-1} = f_{n-1}$;

$g_{n+p}(a_0, a_1, \ldots, a_n, x_{n+1}, \ldots, x_{n+p}) = x$ for $p \in \mathbf{N}$

$g_{n+p}(z_0, z_1, \ldots, z_{n+p}) = f_{n+p}(z_0, z_1, \ldots, z_{n+p})$ if

$(z_0, z_1, \ldots, z_n) \neq (a_0, a_1, \ldots, a_n)$

The normal algorithm G accelerates

$$S \cup \{(a_0, a_1, \ldots, a_n, x_{n+1}, \ldots)\}$$

with the velocity (ε_n) and hence S is not maximal.

REFERENCES

[1]   BREZINSKI C.
        "Accélération de la convergence en Analyse Numérique"
        Lecture Notes in Mathematics, 584, Springer-Verlag,
        Heidelberg, 1977.

[2]   BREZINSKI C.
        "Algorithmes d'accélération de la convergence :
        Etude Numérique"
        Technip, Paris, 1978.

[3]   DELAHAYE J.P.
        "Liens entre la suite du rapport des erreurs et celle
        du rapport des différences : démonstrations"
        Publication A.N.O., n° 14, Université des Sciences te
        Techniques de Lille, 1979.

[4]   DELAHAYE J.P.
        "Liens entre la suite du rapport des erreurs et celle du
        rapport des différences"
        C.R. Acad. SDc. Paris, 290 A, 1980, pp. 343-346.

[5]   DELAHAYE J.P.
        "Optimatité du procédé Δ^2 d'Aitken pour l'accélération de
        la convergence linéaire"
        RAIRO, Analyse Numérique, 15, 1981, pp. 321-330

[6]   DELAHAYE J.P.
        "The partially ordered system of accelerable families"
        85th Summer Meeting of the A.M.S.,Pittsburgh (U.S.A.), Août
        1981.

[7]   DELAHAYE J.P.
        "Les divers types de transformations algorithmiques"
        Publication ANO n° 69, Université des Sciences et Techniques
        de Lille, 1982.

[8]   DELAHAYE J.P.
        "Systèmes ordonnés de familles accélérables"
        Publication ANO n° 73, Université des Sciences et Techniques
        de Lille, 1982.

[9]   DELAHAYE J.P. et GERMAIN-BONNE B.
        "Résultats négatifs en accélération de la convergence"
        Numer. Math., 35, 1980, pp. 443-457.

[10]   DELAHAYE J.P. et GERMAIN-BONNE B.
        "The set of logarithmically convergent sequences cannot be
        accelerated"
        SIAM Num. Anal. 19, 1982, pp. 840-844.

[11]   GERMAIN-BONNE B.
        "Estimation de la limite de suites et formalisation des
        procédés d'accélération de convergence"
        Thèse d'Etat, Lille, 1978.

[12]   GERMAIN-BONNE B.
        "Conditions suffisantes d'accélérabilité"
        Publications ANO n° 32, Université des Sciences et Techniques
        de Lille, 1981.

[13]   GERMAIN-BONNE B.
        Communication personnelle, 1981.

[14]   WIMP J.
        "Sequence transformations and their applications"
        Academic Press, New York, 1981.

Chapter 5

Non-Accelerable Families
of Sequences

INTRODUCTION

Let **S** be a family of convergent sequences. The simplest acceleration problem for **S** is : Is there an (algorithmic or normal, etc...) transformation accelerating **S** ?

Two types of answers may be given :

1°) Positive answers : we take a well-known transformation (or build a new one), then show that it accelerates all the sequences in **S**. Such results have been given about most existing acceleration methods ([3],[4],[5],[25],...) .

2°) Negative answers : thanks to a reduction to absurdity, considering more or less general types of sequence transformations, to have a transformation accelerating every sequence of **S** is established as being impossible. These results show either that we have to take a bigger family of transformations (some results of Pennacchi [23] and Germain-Bonne may be viewed in this way) or have to renounce to accelerate entirely **S**, and that we have to try to accelerate subfamilies of **S** (since, the type of transformations is very general here, this is the meaning of our results).

In this chapter, we only deal with non-accelerable families and with the proofs of non-accelerability results.

The remanence is a very efficient tool to prove such results. Indeed, this sufficient condition of non-accelerability applies to numerous families. In section 1, the remanence is presented in a detailed manner, with variants and easy applications.

Monotone sequences seem to be easily accelerated, but in fact we prove, in section 2, that numerous families of monotone sequences (even some rather small ones) are not accelerable.
The remanence is used except in the proof of the latter theorem, where a direct reasoning is necessary.

In section 3, we are concerned with oscillating and alternating sequences. We do not obtain any more necessary and sufficient conditions of accelerability, for the situation is now too complicated. Remanence is used once and a direct proof is also used once.

In section 4, we study linearly convergent sequences and, in section 5, we study logarithmically convergent sequences.

The methods of proving whether certain families are remanent or not are based upon a technique analogous to the diagonalisation used in theory of cardinality and in mathematical logic. These efficient methods certainly will give rise to many other results about algorithms in numerical analysis (see [24]), in particular in optimization.

Most of the results of this chapter have been presented in the author's articles ([9],[10]), or in articles written with B. GERMAIN-BONNE ([13],[14],[15] and [16]). However, the presentation here is new and more systematic.

Notation

E   : metric space with distance d ;

$E^{(N)}$: set of finite sequences of elements of E ;

E^N : set of infinite sequences of elements of E ;

Conv(E) : set of convergent sequences of E ;

$\qquad$ if $(x_n) \in$ Conv(E) , we write $x = \lim\limits_{n \to \infty} x_n$

$Conv^*(E)$: $\{(x_n) \in$ Conv(E) $\mid \exists\, n_0 \in N , \forall\, n \geq n_0 , x_n \neq x\}$

E^α : set of accumulation points of E ;

$\qquad$ $E^\alpha = \{x \in E \mid \forall\, \epsilon > 0 , \exists\, y \in E : 0 < d(x,y) \leq \epsilon\}$

If $E \subset R$,

E^γ : set of left accumulation points of E

$\qquad$ $E^\gamma = \{x \in E \mid \forall\, \epsilon > 0 , \exists\, y \in E , x - \epsilon < y < x\}$;

E^δ : set of right accumulation points of E

$\qquad$ $E^\delta = \{x \in E \mid \forall\, \epsilon > 0 , \exists\, y \in E , x < y < x+\epsilon\}$;

$^\tau Norm_k(E,F)$: set of normal transformations from E into F
$\qquad$ (see ch. 1)

$^\tau Alg(E,F)$: set of algorithmic transformations from E into F
$\qquad$ (see ch. 1)

1 — REMANENCE AND FIRST APPLICATIONS

As elaborated in [13], [15] and [16] , the remanence is now the main tool to prove negative results in convergence acceleration. This sufficient condition of non-accelerability is presented here as three different forms, called generalized remanence (GR) , remanence (R) and restricted remanence (RR) .

The generalized remanence is the most important one, for it is the most easily satisfied (proposition 1). However, (GR) is rather complicated and consequently is less convenient to use than (R) and (RR).

The main result of course is theorem 1, which states that every family satisfying (GR) is non-accelerable and gives analogous corollaries for (R) and (RR). The given proof is new, but can be obtained through generalizing the proof of [15] concerning (R). The remarks following this proof are essential and perhaps need further developments.

Propositions 2 and 3 and theorem 4 and 5 are direct applications of theorem 1, and allow us to obtain our first non-accelerable families (some of them are simple and relatively small).

Definitions

Let $S \subset \text{Conv}(E)$.

We say that S is remanent in the general sense, if S satisfies the following condition (called generalized remanence)

(GR)

(a) There exists $(x^n) \in \text{Conv}^*(E)$ such that :

(0°) there exists $(x^0_n) \in S$ such that $(x^0_n) \to x^0$

(1°) for every $m_0 \geq 0$, There exist $p_0 \geq m_0$ and

$(x^1_n) \in S$ such that $(x^1_n) \to x^1$ and :

$\forall\, m \leq p_0$, $x^1_m = x^0_m$;

(2°) for every $m_1 > p_0$, there exist $p_1 \geq m_1$ and

$(x^2_n) \in S$ such that $(x^2_n) \to x^2$ and :

$\forall\, m \leq p_1$: $x^2_m = x^1_m$;

...   .

(i°) for every $m_{i-1} > p_{i-2}$ there exist $p_{i-1} \geq m_{i-1}$

and $(x^i_n) \in S$ such that $(x^i_n) \to x^i$ and :

$\forall\, m \leq p_{i-1}$: $x^i_m = x^{i-1}_m$;

...   .

(b) $(x^0_0, x^0_1, \ldots, x^0_{p_0}, x^1_{p_0+1}, \ldots, x^1_{p_1}, x^2_{p_1+1}, \ldots, x^{i-1}_{p_{i-1}},$

$x^i_{p_{i-1}+1}, \ldots, x^i_{p_i}, x^{i+1}_{p_i+1}, \ldots) \in S$

We say that S is remanent if S satisfies the following relation, (called remanence) :

(R)

(a)   there exists $(x^n) \in \mathrm{Conv}^*(E)$ such that :

$(0°)$ there exists $(x^0{}_n) \in S$ such that $(x^0{}_n) \to x^0$;

$(1°)$ for every $m_0 \geq 0$,   there exists $(x^1{}_n) \in S$

such that $(x^1{}_n) \to x^1$ and :

$\forall\ m \leq m_0 :\ x^1{}_m = x^0{}_m$;

$(2°)$ for every $m_1 > m_0$,   there exists

$(x^2{}_m) \in S$ such that $(x^2{}_n) \to x^2$ and :

$\forall\ m \leq m_1 :\ x^2{}_m = x^1{}_m$;

...   .

$(i°)$ for every $m_{i-1} > m_{i-2}$,   there exists $(x^i{}_n) \in S$

such that $(x^i{}_n) \to x^i$ and   :

$\forall\ m \leq m_{i-1}$:

$x^i{}_m = x^{i-1}{}_m$;

...   .

(b) $(x^0{}_0, x^0{}_1, \ldots, x^0{}_{m_0}, x^1{}_{m_0+1}, \ldots, x^1{}_{m_1}, x^2{}_{m_1+1}, \ldots, x^{i-1}{}_{m_{i-1}})$

$x^i{}_{m_{i-1}+1}, \ldots, x^i{}_{m_i}, x^{i+1}{}_{m_i+1}, \ldots) \in S$

We say that **S** is remanent in a restricted sense, if **S** satisfies the following conditions, (called restricted remanence) :

(RR)

(a)   there exists $(x^n) \in \text{Conv}^*(E)$;

(b)   there exist $(x^0{}_n), (x^1{}_n), \ldots, (x^i{}_n), \ldots$,

such that

$(x^0{}_n) \to x^0, (x^1{}_n) \to x^1, \ldots, (x^i{}_n) \to x^i, \ldots$;

(c)   for every strictly increasing sequence of integers (m_i) :

$(x^0{}_0, x^0{}_1, \ldots, x^0{}_n, \ldots) \in \mathbf{S}$

$(x^0{}_0, x^0{}_1, \ldots, x^0{}_{m_0}, x^1{}_{m_0+1}, \ldots, x^1{}_m, \ldots) \in \mathbf{S}$;

$\ldots \quad \ldots \quad \ldots$

$(x^0{}_0, x^0{}_1, \ldots, x^0{}_{m_0}, x^1{}_{m_0+1}, \ldots, x^1{}_{m_1}, x^2{}_{m_1+1}, \ldots, x^{i-1}{}_{m_{i-1}},$

$x^i{}_{m_{i-1}+1}, \ldots, x^i{}_m, \ldots) \in \mathbf{S},$

$\ldots \quad \ldots \quad \ldots$

(d) $(x^0{}_0, x^0{}_1, \ldots, x^0{}_{m_0}, x^1{}_{m_0+1}, \ldots, x^1{}_{m_1}, x^2{}_{m_1+1}, \ldots, x^{i-1}{}_{m_{i-1}}$

$x^i{}_{m_{i-1}+1}, \ldots, x^i{}_{m_i}, x^{i+1}{}_{m_i+1}, \ldots) \in \mathbf{S}$.

Remark

It is well known that :

$$[\exists a \ \forall b \ P(a,b)] \Rightarrow [\forall b \ \exists a \ P(a,b)]$$

From (R) to (RR) , there are only such permutations of quantifiers.

Proposition 1

$(RR) \Rightarrow (R) \Rightarrow (GR)$.

Proof

"(RR) => (R)" : the sequence (x^i_n) in (R) is obtained from the sequence (x_n) in (RR) and (R(c)).

"(R) => (GR)" : we choose $p_i = m_i$.

Theorem 1

If the family $S \subset Conv(E)$ satisfies (GR) , then there is no algorithmic transformation accelerating S.

With proposition 1, we obtain :

Theorem 2

If the family $S \subset Conv(E)$ satisfies (R) , then there is no algorithmic transformation accelerating S.

Theorem 3

If the family $S \subset Conv(E)$ satisfies (RR) , then there is no algorithmic transformation accelerating S.

Proof of theorem 1

Let $\lambda \in \mathbf{R}$, $0 < \lambda < 1$.

Let $S \subset Conv(E)$ satisfy (GR) .

Suppose there exists an algorithm for sequences $A = (\mathbf{R}, \mathbf{C})$ which accelerates the convergence of every sequence of S.

Let $(x^n) \in Conv^*(E)$ be obtained from (GR(a)).

Step 0

Let $(x^0_n) \in S$ be obtained from $(GR(a)(0^o))$. A transforms (x^0_n) into a sequence (t^0_n) accelerating the convergence of (x^0_n) and thus (t^0_n) has the same limit as (x^0_n).

If $x^0 = x$, we set $m_0 = 0$;

If $x^0 \neq x$, we determine $n_0 \geq 0$ such that :

$$d(t^0_{n_0}, x)/d(x^0_{n_0}, x) > \lambda \quad \text{and} \quad \forall n \geq n_0 : \quad d(x^0_n, x^0) \leq 1/2^0 .$$

The calculation of $t^0_0, t^0_1, \ldots, t^0_{n_0}$ only needs a finite number

of points of (x^0_n). Let $m_0 \geq n_0$ be greater than the biggest index of used points. For every sequence (x_n) begining with the same m_0+1 first points as (x^0_n), the algorithm A gives the same n_0+1 first answers $t_0, t_1, \ldots, t_{n_0}$, and thus :

$$d(t_{n_0}, x) / d(x_{n_0}, x) > \lambda \qquad (P_0)$$

... ...

Step i

Let $p_{i-1} > m_{i-1}$ and $(x^i_n) \in S$ be obtained from $(GR(a)(i^0))$. A transforms (x^i_n) into a sequence (t^i_n) accelerating the convergence of (x^i_n), and thus (t^i_n) has the same limit as (x^i_n). If $x^i = x$, we set $m_i = p_{i-1}+1$.

If $x^i \neq x$, we determine $n_i > n_{i-1}$ such that :

$$d(t^i_{n_i}, x) / d(x^i_{n_i}, x) > \lambda \quad \text{and} \quad \forall n \geq n_i : d(x^i_n, x^i) \leq 1/2^i$$

The calculation of $t^i_0, t^i_1, \ldots, t^i_{n_i}$ only need a finite number of

points of (x^i_n). Let $m_i > p_{i-1}$, $m_i > n_i$ be greater than the biggest index of used points. For every sequence (x_n) begining with the same m_i+1 first points as (x^i_n), the algorithm A gives the same n_i+1 first answers $t_0, t_1, \ldots, t_{n_i}$, and thus :

$$d(t_{n_i}, x) / d(x_{n_i}, x) > \lambda \qquad (P_i)$$

... ...

Now let (x_n) be obtained from $(GR(b))$. By construction, $(x_n) \to x$. (P_i) is satisfied for i sufficiently large (for $x^i \neq x$ when i sufficiently large). Consequently, A does not accelerate (x_n).

Remarks

1°) Note that in theorem 1, 2 and 3, we say "then there is no algorithmic transformation accelerating S" and not only "then S is not accelerable" (which means "there is no normal algorithm accelerating S"). Which is a surprising fact, for the definition of an algorithmic transformation allows t_n to depend upon points x_i with $i \geq n$: even if t_n is obtained from $x_0, x_1, \ldots, x_{2n}$ (or $x_0, x_1, \ldots, x_{n^2}$ or $x_0, x_1, \ldots, x_{n^n}$) the sequence (t_n) does not accelerate (x_n) for every $(x_n) \in S$ when S is remanent.

For certain difficult families of sequences this is no longer the case and in order to prove non-accelerability an entirely new proof is necessary.   Our attempts to obtain a necessary and sufficient condition of non-accelerability were not succesful, and (GR), (R) and (RR)   remain sufficient but far from necessary conditions of non-accelerability.

2°) The relations $(P_0),\ldots,(P_i),\ldots$ do not only mean that (x_n) is not accelerated by (t_n) , but mean that (x_n) is not improved by (t_n) , if we define improvement by the following.   The sequence (t_n) improves the convergence of (x_n) by λ if :

$$\exists\ n_0 \in \mathbf{N}\ ,\ \forall\ n \geq n_0\ ,\ d(t_n,x) \leq \lambda\ d(x_n,x)\ .$$

It is impossible to improve by λ the convergence of any family (GR) satisfying (GR), whatever $\lambda \in \mathbf{R}$, $0 < \lambda < 1$ may be.

3°) The hypothesis that A accelerates the convergence of S is not really utilized. In fact, we only need that   A   is regular for $(x^0{}_n),\ldots,(x^i{}_n),\ldots$ and accelerative for (x_n) .

Consequently, we can give the following more general result :

If   (S',S)   satisfy   (GR')   then there is no algorithmic transformations   A $\in$ $\mathbf{Alg}(E,E)$ regular on   S'   and accelerative on   S. (GR')   is the relation obtained from   (GR)   by substituting   S   for S'   in   (RG(a)).

Analogous results are obtained from   (R)   and   (RR).

This agrees with the fact that   "the size of the domain of regularity of a transformation and its efficiency seem to   be inversely related" (Jet Wimp [25]).

From a convergent sequence $(x_n) \in \mathbf{Conv}^*(E)$, by "stretching" one can obtain very slowly convergent sequences.   For exemple :

$$(y_n) = (x_0,x_0,x_1,x_1,x_2,x_2,\ldots,x_i,x_i,x_{i+1},x_{i+1},\ldots)$$

$$(z_n) = (x_0,x_1,x_1,x_2,x_2,x_2,x_3,x_3,x_3,x_3,x_4,\ldots)$$

With this method, we shall construct our first non-accelerable family

Definition

Let $(x_n) \in E^{\mathbf{N}}$.   We denote by $\mathbf{Stret}(x_n)$ the set of "stretching sequences" obtained from (x_n) , that is of the form $(y_n) = (x_{i_n})$ where (i_n) is a sequence of integers satisfying :

$$\forall\ n \in \mathbf{N}\ \ i_{n+1} = i_n\ \ \text{or}\ \ i_n+1.$$

Proposition 2

If $(x_n) \in Conv^*(E)$, then $Stret(x_n)$ satisfies (RR).

Proof

In (RR(a)), we choose $(x^n) = (x_n)$

In (RR(b)), we choose :

$$(x^0_n) = (x_0, x_0, \ldots, x_0, \ldots)$$

$$\ldots \quad \ldots \quad \ldots$$

$$(x^i_n) = (x_i, x_i, \ldots, x_i, \ldots)$$

$$\ldots \quad \ldots \quad \ldots$$

Corollary

Two accelerable families may have a non-accelerable union.

Proof

Let $(x_n) \in Conv^*(E)$.

We define :

$$StretA(x_n) = \{(x_{i_n}) \mid i_{n+1} = i_n \text{ or } i_n+1 \text{ and } \lim_{n \to \infty} i_n = + \infty\}$$

$$StretB(x_n) = \{(x_{i_n}) \mid i_{n+1} = i_n \text{ or } i_n+1 \text{ and } \exists N , \forall n \geq N : i_n \leq N\}$$

The transformation defined by $t_n = x$ accelerates $StretA(x_n)$ and the transformation defined by $t_n = x_n$ accelerates $StretB(x_n)$, although $Stret(x_n) = StretA(x_n) \cup StretB(x_n)$ is not accelerable.

Proposition 2 leads us to the following general theorem, which gives a necessary and sufficient condition of the existence of a universal algorithm on a set E (i.e. an algorithm which accelerates $Conv(E)$). This condition is very simple and the result means that when E contains numerous elements ($E = \mathbf{R}^n$, $E = [a,b]$) , there is no universal algorithm.

Theorem 4

$Conv(E)$ is accelerable $\iff E^\alpha = \emptyset$.

Proof

The condition is sufficient, for if $E^\alpha = \emptyset$, then $\mathrm{Conv}(E) \subset \mathrm{UStat}(E)$ which is accelerated, (even predicted) by $t_n = x_n$.

The condition is also necessary, for if $E^\alpha \neq \emptyset$, then there exists $(x_n) \in \mathrm{Conv}^*(E)$ and $\mathrm{Stret}(x_n) \subset \mathrm{Conv}(E)$.

It is also of interest to know under what conditions $\mathrm{Conv}^*(E)$ is accelerable. Since $\mathrm{Conv}^*(E)$ is included in $\mathrm{Conv}(E)$, we must find that $\mathrm{Conv}^*(E)$ is more often accelerable than $\mathrm{Conv}(E)$. Theorem 5 shows this to be true. Similar to the notion of stretching, we introduce the notion of diagonalisation.

Definition

Let $(x^0{}_n),\ldots,(x^i{}_n),\ldots$ be in E^N . We denote by

$$\mathrm{Dia}((x^0{}_n),\ldots,(x^i{}_n),\ldots)$$

the   set   of   sequences   obtained   by   diagonalisation   from $(x^0{}_n),\ldots,(x^i{}_n),\ldots,$ that is of the form $(y_n) = (x\,{}^{i_n}{}_n)$,   where (i_n) is a sequence of integers satisfying :

$$\forall\, n \in N \quad i_{n+1} = i_n \quad \text{or} \quad i_n+1.$$

Proposition 3

Let $(x^0{}_n),\ldots,(x^i{}_n)$ be in $\mathrm{Conv}(E)$.

If $(x^n) \in \mathrm{Conv}^*(E)$ and if $\mathrm{Dia}((x^0{}_n),\ldots,(x^i{}_n),\ldots) \subset \mathrm{Conv}(E)$ then $\mathrm{Dia}((x^0{}_n),\ldots,(x^i{}_n),\ldots)$ satisfies (RR).

Proof

In (RR(a)), we choose (x^n) .

In (RR(b)) , we choose $(x^0{}_n),\ldots,(x^i{}_n),\ldots$.

Theorem 5

$\mathrm{Conv}^*(E)$ is accelerable $\iff$ $(E^\alpha)^\alpha = \emptyset$.

Proof

(a) Assume that $(E^\alpha)^\alpha = \emptyset$.

For every $n \in \mathbf{N}$, let $f_n : E \to E$ be a function satisfying :

$$\forall\, t \in E : f_n(t) \in E^\alpha \quad \text{and } d(t, f_n(t)) \leq d(t, E^\alpha) + 1/n \ .$$

The family (f_n) is a 1-memory-algorithm which accelerates (even predicts) every sequence of $Conv^*(E)$.

Indeed, if $(x_n) \in Conv^*(E)$, then $x \in E^\alpha$, $d(x, E^\alpha - \{x\}) > 0$ and thus :

$$\forall\, n > 2/d(x, E^\alpha - \{x\}) : f_n(x_n) = x \ .$$

(b) Assume that $(E^\alpha)^\alpha \neq \emptyset$.

Let $(x^n) \in Conv^*(E)$ be such that $x \in (E^\alpha)^\alpha$ and such that :

$$\forall\, n \in E \ , \quad x^n \in E^\alpha \ .$$

For every $i \in \mathbf{N}$, consider $(x^i_n) \in Conv^*(E)$ such that :

$$\forall\, n \in \mathbf{N}, \ x^i_n \neq x^i \quad \text{and } d(x^i_n, x^i) \leq 1/(i+1) \ .$$

We obtain that $Dia((x^0_n), \ldots, (x^i_n,), \ldots) \subset Conv^*(E)$, thus $Conv^*(E)$ is not accelerable (proposition 3).

Remark

If one algorithm $A \in Alg(E,E)$ is not sufficient to accelerate $\mathbf{S}$, it is possible that two algorithms $A_1, A_2 \in Alg(E,E)$ can accelerate $\mathbf{S}$. (A_1 would accelerate $\mathbf{S}_1$, A_2 would accelerate $\mathbf{S}_2$ and $\mathbf{S}_1 \cup \mathbf{S}_2 = \mathbf{S}$). Concerning $Conv^*(E)$ $(E = \mathbf{R})$, in $[15]$ we show (with some restrictions on the algorithms), that two algorithms are not sufficient, nor any denumerable number of algorithms.

2 – FAMILIES OF MONOTONE SEQUENCES

After $Conv(E)$ and $Conv^*(E)$, we naturally study the families of monotone sequences. As in § 1 , since the situation is still rather simple, we obtain necessary and sufficient conditions of accelerability (theorems 6, 7). The negative part of these results is directly obtained from remanence.

The families considered in theorem 8 are smaller, which is why we only obtain a sufficient condition of non-accelerability (not given by remanence).

Definitions, notation

Let $E \subset \mathbf{R}$.

We denote by $\text{Mon}^+(E)$ the family of increasing convergent sequences of E :

$$\text{Mon}^+(E) = \{(x_n) \in \text{Conv}(E) \mid \forall\, n \in \mathbf{N} : x_{n+1} \geq x_n\} ;$$

* the family of decreasing convergent sequences :

$$\text{Mont}^-(E) = \{(x_n) \in \text{Conv}(E) \mid \forall\, n \in \mathbf{N} : x_{n+1} \leq x_n\} ;$$

* the family of monotone convergent sequences :

$$\text{Mon}(E) = \text{Mon}^+(E) \cup \text{Mon}^-(E) ;$$

* the family of strictly increasing convergent sequences :

$$\text{Mon}^{+*}(E) = \{(x_n) \in \text{Conv}(E) \mid \forall\, n \in \mathbf{N} : x_{n+1} > x_n\} ;$$

* Analogously, we define :

$$\text{Mon}^{-*}(E) ; \text{Mon}^*(E) ;$$

* the family of ultimately increasing convergent sequences :

$$\text{UMon}^+(E) = \{(x_n) \in \text{Conv}(E) \mid \exists\, n_0 \in \mathbf{N} , \forall\, n \geq n_0 : x_{n+1} \geq x_n\} ;$$

* Analogously, we define :

$$\text{UMon}^-(E) ; \text{UMon}(E) ; \text{UMon}^{+*}(E) ; \text{UMon}^{-*}(E) ; \text{UMon}^*(E) ;$$

* the family of k-increasing sequences $(k \in \mathbf{N}^*)$:

$$\text{Mon}^+_k(E) = \{(x_n) \in \text{Conv}(E) \mid \forall\, i \in \{1,\ldots,k\} \,\forall\, n \in \mathbf{N}$$
$$(-1)^i\, \Delta^i\, x_n \leq 0\}$$

* Analogously, we define :

$$\text{Mon}^-_k(E) ; \text{Mon}_k(E) ; \text{Mon}^{+*}_k(E) ; \text{Mon}^{-*}_k(E) ; \text{Mon}^*_k(E) .$$

There are lots of relationships between these families. For example :

$$\text{Mon}^{*+} \subset \text{Mon}^* \subset \text{Mon} \subset \text{UMon} .$$

$$\text{Mon}_{k+1} \subset \text{Mon}_k \subset \text{Mon}_1 = \text{Mon} .$$

Theorem 6

(i) $Mon^-(E)$ is accelerable $<=> E^\delta = \emptyset$;

(ii) $Mon^+(E)$ is accelerable $<=> E^\gamma = \emptyset$;

(iii) $Mon(E)$ is accelerable $<=> E^\alpha = \emptyset$.

Proof

(i)   If $E^\delta = \emptyset$, every decreasing convergent sequence of
      E^N is ultimately stationary and thus is accelerated (in
      fact predicted) by $t_n = x_n$.

      If $E^\delta \neq \emptyset$, there exists $(x^n) \in Mon^{-*}(E)$ with
      $x \in E^\delta$. Since $Stret(x^n) \subset Mon^-(E)$ from proposition 1,
      we obtain that $Mon^-(E)$ is not accelerable.

(ii) Similar to (i)

(iii) Easily obtained from (i) and (ii).

Theorem 7

(i) $Mon^{-*}(E)$ is accelerable $<=> (E^\delta)^\delta = \emptyset$;

(ii) $Mon^{+*}(E)$ is accelerable $<=> (E^\gamma)^\gamma = \emptyset$;

(iii) $Mon^*(E)$ is accelerable $<=> (E^\delta)^\delta \cup (E^\gamma)^\gamma = \emptyset$.

Proof

(i)   If $(E^\delta)^\delta = \emptyset$, we define the following 1-stationary
      algorithm :

$$f(x_n) = \max\{t \in E^\delta \mid t \leq x_n\} ,$$

It is easy to verify that this algorithm accelerates (in
fact predicts) $Mon^{-*}(E)$.

If $(E^\delta)^\delta \neq \emptyset$, there exists $(x^n) \in Mon^{-*}(E^\delta)$ with
limit $x \in (E^\delta)^\delta$ and for every $i \in N$, there exists
$(x^i_n) \in Mon^{-*}(E)$ such that :

$$\forall n \in N , \quad x^i < x^i_n < x^{i+1} ;$$

It is easy to verify that $Mon^{-*}(E) \supset Dia((x^o_n),...,(x^i_n),...)$

Thus (from proposition 3), $Mon^{-*}(E)$ is not accelerable.

(ii) Similar, to (i).

(iii) If $(E^\delta)^\delta \cup (E^\gamma)^\gamma \neq \emptyset$, from (i) and (ii) we obtain
       that $\text{Mon}^*(E)$ is not accelerable.

       If $(E^\delta)^\delta \cup (E^\gamma)^\gamma = \emptyset$, we define the following 2-stationary
       algorithm :

   $$f(x_{n-1},x_n) = \max\{t \in E^\delta \mid t \leq x_n\} \quad \text{if} \quad x_n < x_{n-1}$$

   $$f(x_{n-1},x_n) = \min\{t \in E^\gamma \mid t \geq x_n\} \quad \text{if} \quad x_n > x_{n-1} .$$

Remarks

$1°$) It is possible to simultaneously have $(E^\alpha)^\alpha \neq \emptyset$ and
       $(E^\delta)^\delta \cup (E^\gamma)^\gamma = \emptyset$. This is the case with :

   $E = \{1 - 1/2^n + 1/2^m \mid n,m \in \mathbb{N} \;\; n+1 < m\} \cup \{1 - 1/2^n \mid n \in \mathbb{N}\} \cup \{1\}$.

$2°$) Theorems 6 and 7 are still true for :

   $\text{UMon}^+(E)$, $\text{UMon}^-(E)$, $\text{UMon}(E)$, $\text{UMon}^{+*}(E)$, $\text{UMon}^{-*}(E)$,

   $\text{UMon}^*(E)$.

$3°$) Note that 1-stationary and 2-stationary algorithms defined in
     the proof of theorem 7 are not regular on $\text{Conv}(E)$: again good
     properties of accelerability are obtained at the expense of
     properties of regularity.

Theorem 8

Let $k \in \mathbb{N}^*$. Let $E \subset \mathbb{R}$ satisfy $\overset{\circ}{E} \neq \emptyset$.

None of the following families are accelerable :

   $\text{Mon}^{+*}_k(E)$, $\text{Mon}^+_k(E)$, Mon^{-*}_k , $\text{Mon}^-_k(E)$, $\text{Mon}^*_k(E)$, $\text{Mon}_k(E)$.

Lemma 1

For every $k \in \mathbb{N}^*$, there exists $\varepsilon_k \in \mathbb{R}^{+*}$ such that if (x_n)
satisfies :

   $$\left| \begin{array}{l} \forall \, n \in \mathbb{N} : (x_{n+2} - x_{n+1}) \,/\, (x_{n+1} - x_n) = 1/2 \quad \text{or} \quad (1 - \varepsilon_k)/2 , \\ x_1 > x_0 , \end{array} \right.$$

then $(x_n) \in \text{Mon}^{+*}_k(\mathbb{R})$.

Proof

Let (x_n) be a sequence such that :

$$\forall \, n \in \mathbb{N} : (x_{n+2} - x_{n+1}) \, / \, (x_{n+1} - x_n) = 1/2 \quad \text{or} \quad (1-\epsilon)/2$$

$$x_1 > x_0$$

By induction on i , we will show that :

$$\forall \, i \in \mathbb{N} \, , \, \forall \, n \in \mathbb{N} : \Delta^i x_n = (x_{n+1} - x_n) \; f_{i,J}(\epsilon)(-1)^{i+1}$$

where J is the set of $j \in \{1,\dots,i-1\}$ satisfying :

$$(x_{n+j+1} - x_{n+j}) \, / \, (x_{n+j} - x_{n+j-1}) = (1-\epsilon)/2 \, ,$$

and $f_{i,J}$ is a function continuous at 0 and such that :

$$f_{i,J}(0) = \frac{1}{2^{i-1}} > 0$$

(for example : $f_{1,\emptyset}(\epsilon) = 1$; $f_{2,\emptyset}(\epsilon) = 1/2$;

$$f_{2,\{1\}}(\epsilon) = 1 - (1-\epsilon)/2 \; ; \; f_{3,0}(\epsilon) = 1/4 \; ;$$

$$f_{3,\{1\}}(\epsilon) = 1 - 2(\frac{1-\epsilon}{2}) + \frac{1}{2}(\frac{1-\epsilon}{2}) \; ;$$

$$f_{3\{4\}}(\epsilon) = 1 - 1 + \frac{1}{2}(\frac{1-\epsilon}{2}) \; ;$$

$$f_{3,\{1,2\}}(\epsilon) = 1 - 2(\frac{1-\epsilon}{2}) + (\frac{1-\epsilon}{2})^2)$$

Thus, there exists $\epsilon_k \in \mathbb{R}^{+*}$ such that :

$$\forall \, n \in \mathbb{N} : x_{n+1} > x_n \, ,$$

$$\forall \, i \subset \{1,2,\dots,k\} \, , \, \forall \, J \subset \{1,\dots,i-1\} \; f_{i,J}(\epsilon_k) > 0 \, .$$

With this ϵ_k , we obtain :

$$\forall \, n \in \mathbb{N} \, , \, \forall \, i \in \{1,2,\dots,k\} : (-1)^i \Delta^i x_n < 0 \, .$$

Thus $(x_n) \in \text{Mon}^{+*}{}_k(\mathbb{R})$.

Lemma 2

Let $\varepsilon \in R^{+*}$, $\varepsilon < 1$.

The family of sequences (x_n) such that :

(S_ε) $\left|\begin{array}{l} \forall\ n \in \mathbf{N} : (x_{n+2} - x_{n-1}) / (x_{n+1} - x_n) = 1/2 \quad \text{or} \quad (1-\varepsilon)/2 \ , \\ x_1 > x_0 \ , \end{array}\right.$

it not accelerable.

Proof

Let   T   be a normal transformation which accelerates the family of sequences satisfying (S_ε) .

Let $(x^0{}_n) = -1/2^n$.

From the hypothesis, there exists n_0 such that :

$$(t^0{}_{n_0} - x^0) / (x^0{}_{n_0} - x^0) \leq \varepsilon/2.$$

For every $x \in {}_0[x^0{}_n \ , \ x^1]$ where $x^1 = x^0 - \varepsilon (x^0 {}_0 - x^0{}_n)$,

we have :

$$(t^0{}_{n_0} - x) / (x^0{}_{n_0} - x) \geq \varepsilon/2\ (1 - \varepsilon) > 0. \qquad (A_0)$$

Let $(x^1{}_n)$ be defined by :

$x^1{}_n = x^0{}_n$ if $n \leq n_0$,

$(x^1{}_{n_0+1} - x^1{}_{n_0}) / (x^1{}_{n_0} - x^1{}_{n_0-1}) = (1 - \varepsilon)/2$,

$(x^1{}_{n+1} - x^1{}_n) / (x^1{}_n - x^1{}_{n-1}) = 1/2$ if $n \geq n_0+1$.

This sequence $(x^1{}_n)$ converges to x^1.

From the hypothesis, there exists $n_1 > n_0$ such that :

$$(t^1{}_{n_1} - x^1) / (x^1{}_{n_1} - x^1) \leq \varepsilon/2 \ .$$

For every $x \in [x^1{}_{n_1}, x^2]$ where $x^2 = x^1 - \varepsilon(x^1 - x^1{}_{n_1})$,   we have

$$(t^1{}_{n_1} - x)/(x^1{}_{n_1} - x) \geq \varepsilon/2(1-\varepsilon) > 0 \qquad (A_1)$$

etc...

Then, we define :

$$(x_n) = (x^0{}_0, x^0{}_1, \ldots, x^0{}_{n_0}, x^1{}_{n_0+1}, \ldots, x^1{}_{n_1}, x^1{}_{n_1+1}, \ldots)$$

The sequence (x_n) is increasing and converges to $x \in \cap_i \left[x^i{}_{n_i} - x^{i+1} \right]$

Thus $(A_0), \ldots, (A_i), \ldots$ are true, so T does not accelerate (x_n).

Proof of theorem 8

It is sufficient to prove the result for $Mon^{+*}{}_k(E)$ (other considered families are included in $Mon^{+*}{}_k(E)$ or in $Mon^{-*}{}_k(E)$, which can be treated analogously.

For simplicity, we assume that $E = R$. From lemma 1, there exists $\varepsilon_k \in R^{+*}$ such that every sequence satisfying (S_{ε_k}) is in $Mon^{+*}{}_k(R)$; from lemma 2, this family is not accelerable, so that $Mon^{+*}{}_k(R)$ is not accelerable.

Remark

The proof of the theorem does not involve remanence (it is possible that the families considered are not remanent), it is the first time that "normality" is of essential use in a proof.

3 - ALTERNATING AND OSCILLATING SEQUENCES

Alternating and oscillating sequences seem easy to accelerate and really very simple transformations accelerate some of them ([3],[5]). However, among naturally defined families of alternating and oscillating sequences, only a few are accelerable. The information that a sequence is alternating is not sufficient to permit acceleration. The results of this section were presented at the "Colloque annuel d'Analyse Numérique de Belgodère" ([12]).

Definitions

Let $(x_n) \in Conv(R)$.

We consider the following properties :

(ALT.A) : The sequence $(-1)^n (x_{n+1} - x_n)$ has a constant sign.

(ALT.B) : The sequence $(-1)^n (x_{n+1} - x_n)$ is monotone with a constant sign.

(OSC.A) : The sequence $(-1)^n (x_n-x)$ has a constant sign.

(OSC.B) : The sequence $(-1)^n (x_n-x)$ is monotone with a constant sign.

(LIN$^-$) : $\exists\, \ell \in [-1,0]$: $\lim_{n\to\infty}(x_{n+1}-x)/(x_n-x) = \ell$

(LIN^{--}) : $\lim_{n\to\infty} (x_{n+1}-x)/(x_n-x) = -1$

We obtain 6 families of convergent sequences which are respectively denoted by :

Alt.A ; **Alt.B** ; **Osc.A** ; **Osc.B** ; **Lin$^-$** ; **Lin^{--}** .

Proposition

Alt.A $\supset$ **Osc.A** $\supset$ **Alt.B** $\supset$ **Osc.B** $\supset$ **Lin$^-$** $\supset$ **Lin^{--}** .

Proof

The proof is obvious. However, one can remark that there is no equality. Indeed :

The sequence defined by : $x_{n+1} = x_n + (-1)^n/(n^3 + (-1)^n)$
is in **Alt.A** , but not in **Osc.A** ;

The sequence defined by $x_n = x + (-1)^n/(n+1 + 2 \cos(n\pi/2))$
is in **Osc.A** , but not in **Alt.B** ;

The sequence (x_n) defined by : $x_n = x + (-1)^n/(n^2 + (-1)^n)$
is in **Alt.B** , but not in **Osc.B** ;

The sequence (x_n) defined by : $x_n = x + (-1)^n \prod_{i=0}^{n} \lambda_i$
with $\lambda_i = (1/2, 1/3, 1/2, 1/3, \ldots)$ is in **Osc.B** , but not **Lin$^-$** ;

The sequence (x_n) defined by : $x_n = x + (-1)^n \lambda^n$ with
$0 < \lambda < 1$ is in **Lin$^-$** , but not in **Lin^{--}** .

The family **Lin^{--}** is accelerable, for example by the simplest acceleration transformation :

$$t_n = (x_{n+1} + x_n)/2$$

The family **Lin$^-$** is accelerable by the Aitken δ^2 process ([3],[9]) which is not a linear transformation; **Lin$^-$** is also accelerable by other transformations ([3],[17]).

The family of totally oscillating sequences ($[2],[3]$ and $[18]$) is a subfamily of Lin^- ($[4]$) , hence is accelerable.

For the four remaining families, it is not unreasonable to want to accelerate them, however no positive results are known.

About the largest one, the question is easy to answer :

Theorem 9

Alt.A satisfies (GR).

Proof

(a) Let $(x^n) = 1/2^n$

($0°$) $(x^0{}_n)$ is defined by :

$$x^0{}_n = 1 + (-1)^n/2^{n+1}$$

($1°$) Let $m_0 \in \mathbf{N}$. Define p_0 to be the first even integer $\geq m_0$ and define $(x^1{}_n)$ by :

$$x^1{}_n = x^0{}_n \quad \text{if} \quad n \leq p_0 ;$$

$$x^1{}_n = x^1 + (-1)^n/2^{n+2} \quad \text{if} \quad n > p_0$$

($2°$) Let $m_1 > p_0$. Define p_1 to be the first even integer $\geq m_1$ and define $(x^2{}_n)$ by :

$$x^2{}_n = x^1{}_n \quad \text{if} \quad n \leq p_1 ;$$

$$x^2{}_n = x^2 + (-1)^n/2^{n+3} \quad \text{if} \quad n > p_1$$

...

($i°$) Let $m_{i-1} > p_{i-2}$. Define p_{i-1} to be the first even integer $\geq m_{i-1}$ and define $(x^i{}_n)$ by :

$$x^i{}_n = x^{i-1}{}_n \quad \text{if} \quad n \leq p_{i-1}$$

$$x^i{}_n = x^i + (-1)^n/2^{n+i+1} \quad \text{if} \quad n > p_{i-1}$$

...

(b) One verifies that :

$$(x^0{}_0, x^0{}_1, \ldots, x^0{}_{p_0}, x^1{}_{p_0+1}, \ldots, x^2{}_{p_1+1}, \ldots, x^{i-1}{}_{p_{i-1}}, x^i{}_{p_{1-1}+1}, \ldots,$$

$$x^i{}_{p_i}, x^{i+1}{}_{p_i+1}, \ldots) \in \text{Alt.A}$$

Theorem 10

There is no normal transformation which accelerates   Osc.B.

Proof

Let $(f_n) \in$ **Norm**$(\mathbf{R},\mathbf{R})$ be a transformation which accelerate   Osc.B.

We shall construct a sequence $(x_n) \in$ **Osc**.B   not accelerated by (f_n).

Step 0

We choose $x_0, x_1 \in \mathbf{R}$ with $x_1 > x_0$.

...

Step k

We assume that $x_0, x_1, \ldots, x_{2k-2}, x_{2k-1}$ are defined and satisfy :

(a k-1)

$$x_0 < x_2 < \ldots < x_{2k-4} < x_{2k-2} < x_{2k-1} < x_{2k-3} < \ldots < x_3 < x_1 ,$$

(b k-1)

$$L_{k-1} = \{x^* \in \mathbf{R} | (-1)^{n+1}(x_n - x^*) > 0 \quad \text{decreasing for :}$$

$$n = 0, 1, \ldots, 2k-1\} = \left[(x_{2k-2} + x_{2k-1})/2 , x_{2k-1} \right]$$

(c k-1)

$$\forall \ x^* \in L_{k-1} \ : \ (t_{2k-3} - x^*)/(x_{2k-3} - x^*) \geq 1/2$$

There   exist   sequences $(x_n) \in$ **Osc**.B   with   first   terms $(x_0, x_1, \ldots, x_{2k-1})$; hence $(x_0, x_1, \ldots, x_{2k-1}) \in \text{dom} \ f_{2k-1}$; hence $t_{2k-1} = f_{2k-1}(x_0; x_1, \ldots, x_{2k-1})$ is defined.   Two cases are possible :

Case 1

If $t_{2k-1} > (x_{2k-2} + 3x_{2k-1})/4$,   then we set :

$$x_{2k} = (x_{2k-2} + x_{2k-1})/2 , \ x_{2k+1} = (3x_{2k-2} + 5x_{2k-1})/8.$$

Case 2

If $t_{2k-1} < (x_{2k-2} + 3x_{2k-1})/4$,   then we set :

$$x_{2k} = (x_{2k-2} + 7x_{2k-1})/8 , \ x_{2k+1} = (x_{2k-2} + 15x_{2k-1})/16 .$$

It is clear that **(a.k)** is satisfied.

In case 1 or 2 , we obtain :

(b_k) $$L_k = \left[(x_{2k} + x_{2k+1})/2 \ , \ x_{2k+1}\right]$$

In case 1 :

$$\forall \ x^* \in L_k : (t_{2k-1} - x^*)/(x_{2k-1} - x^*) \geq 4/7$$

In case 2 :

$$\forall \ x^* \in L_k : (t_{2k-1} - x^*)/(x_{2k-1} - x^*) \geq 5/3.$$

Hence **(c_k)** is satisfied.

This sequence (x_n) converges to $x \in \bigcap\limits_{k \in \mathbb{N}} L_k$

(in fact $\{x\} = \bigcap\limits_{k \in \mathbb{N}} L_k$)

Since $\forall \ k \in \mathbb{N} : x \in L_k$, we have $(x_n) \in Osc.B$. However, the relations **(c_k)** show that (f_n) does not accelerate (x_n).

Remark

$1°$) In this proof, the impossibility of acceleration does not come from the regularity properties of (f_n) , but has its origins in the fact that Osc.B contains sequences with very different beginings.

$2°$) The sequence (x_n) of the proof satisfies :

$$(x_{n+2} - x_{n+1})/(x_{n+1} - x_n) \in \{-1/2, \ -1/4, \ -1/8\}$$

Thus we can claim that the set of sequences satisfying :

$$\forall \ n \in \mathbb{N} \quad (x_{n+2} - x_{n+1})/(x_{n+1} - x_n) \in \{-1/2, \ -1/4, \ -1/8\},$$

is not accelerable.

4 – FAMILIES OF LINEARLY CONVERGENT SEQUENCES

The family of linearly convergent sequences is one of the biggest families of accelerable sequences. This is shown by establishing that families very near but a little bit bigger than Lin are not accelerable (theorems 11 and 12).

This results were first published in [9] and [10].

Notation and definitions

We recall that for every $\lambda \in \mathbf{R}$, $\lambda \neq 0$, $\lambda \neq 1$:

$$\text{Lin}_\lambda(\mathbf{R}) = \{(x_n) \in \text{Conv}^*(\mathbf{R}) \mid \lim_{n \to \infty}(x_{n+1} - x)/(x_n - x) = \lambda\}$$

$$\text{Lin}(\mathbf{R}) = \cup\{\text{Lin}_\lambda(\mathbf{R}) \mid -1 \leq \lambda < 1 , \lambda \neq 0\}$$

Let $\lambda, \mu \in \mathbf{R}$ $\lambda < \mu$. We define :

$$\text{Lin}_{\lambda, \mu}(\mathbf{R}) = \{(x_n) \in \text{Conv}^*(\mathbf{R}) \mid \exists \, n_0 \in \mathbf{N} , \forall \, n \geq n_0 :$$

$$(x_{n+1} - x) / (x_n - x) \leq \mu)$$

$$\text{Lin}^\Delta_{\lambda, \mu}(\mathbf{R}) = \{(x_n) \in \text{Conv}^*(\mathbf{R}) \mid \exists \, n_0 \in \mathbf{N} , \forall \, n \geq n_0 :$$

$$\lambda \leq (x_{n+2} - x_{n+1})/(x_{n+1} - x_n) \leq \mu\}.$$

$$\text{SLin}(\mathbf{R}) = \{(x_n) \in \text{Conv}^*(\mathbf{R}) \mid \lim_{n \to \infty}(x_{n+1} - x)/(x_n - x) = 0\}$$

(family of super-linearly convergent sequences).

Theorem 11

Let $\lambda, \mu \in \mathbf{R}$ be such that $0 < \lambda < \mu < 1$:

(i)   there is no normal transformation which accelerates $\text{Lin}^\Delta_{\lambda, \mu}(\mathbf{R})$

(ii)   there is no normal transformation which accelerates $\text{Lin}_{\lambda, \mu}(\mathbf{R})$

We need the following obvious result :

Lemma

Let (x_n) be a sequence of real numbers. If there exist $n_0 \in \mathbf{N}$ and $\lambda \in \mathbf{R}$, $|\lambda| < 1$ such that :

$$\forall \, n \geq n_0 : x_{n+1} - x_n = \lambda \, (x_n - x_{n-1}) ,$$

then :

i) $\forall \, k \in \mathbf{N}$ $x_{n_0+k+1} = x_{n_0} + \dfrac{\lambda(1-\lambda^{k+1})}{1-\lambda} (x_{n_0} - x_{n_0-1})$

ii) $\lim_{n \to \infty} x_n = x_{n_0} + \dfrac{\lambda}{1-\lambda} (x_{n_0} - x_{n_0-1})$

Proof of the theorem

(i) We assume there is a normal transformation $A = (f_n)$ which accelerates $Lin^{\Delta}_{\lambda,\mu}(R)$.

We set $\alpha = \dfrac{\mu}{1-\mu} - \dfrac{\lambda}{1-\lambda} > 0$; $\beta = \min\left\{\dfrac{\alpha-\alpha\lambda}{4\lambda} ; 1\right\} > 0$

$\delta = \dfrac{1}{2}\alpha - \dfrac{\lambda\,\beta}{1-\lambda} > 0$; $\Theta = \dfrac{\delta}{\lambda/(1-\lambda)+\alpha/2} > 0$

We define (x^0_n) by :

$x^0_0 = 0$; $x^0_1 > x^0_0$; $\forall n \in \mathbf{N}$: $(x^0_{n+2} - x^0_{n+1})/(x^0_{n+1} - x^0_n) = \lambda$.

By hypothesis, the transformed sequence of (x^0_n) by A , accelerates (x^0_n) , thus there exists $n_0 \in \mathbf{N}$ such that :

$$\left| t^0_{n_0} - x^\lambda_{n_0} \right| / \left| x^0_{n_0} - x^\lambda_{n_0} \right| \leq \beta ,$$

where :
$$x^\lambda_{n_0} = x^0_{n_0} + \dfrac{\lambda}{1-\lambda}(x^0_{n_0} - x^0_{n_0-1}) .$$

(from the lemma : $x^\lambda_{n_0} = \lim_{n\to\infty} x^0_n$)

We define (x^1_n) by :

$$x^1_n = x^0_n \text{ if } n \leq n_0 ,$$

$$(x^0_n - x^1_{n-1})/(x^1_{n-1} - x^1_{n-2}) = \mu \text{ if } n > n_0 .$$

This sequence is increasing, and from the lemma it converges to :

$$x^\mu_{n_0} = x^0_{n_0} + \dfrac{\mu}{1-\mu}(x^0_{n_0} - x^0_{n_0-1}) > x^\lambda_{n_0} .$$

There exists $n_1 > n_0$ such that :

$$x^1_{n_1} \geq x^{\lambda\mu}_{n_0} ,$$

where :

$$x^{\lambda\mu}_{n_0} = (x^\mu_{n_0} + x^\mu_{n_0})/2 = x^\lambda_{n_0} - \alpha(x^0_{n_0} - x^0_{n_0-1})/2$$

We define (x^2_n) by :

$$x^2_n = x^1_n \quad \text{if} \quad n \leq n_1$$

$$(x^2_n - x^2_{n-1})/(x^2_{n-1} - x^2_{n-2}) = \lambda \quad \text{if} \quad n > n_1 .$$

From the hypothesis, the transformed sequence of (x^2_n) by A , accelerates (x^2_n) , thus there exists $n_2 > n_1$, such that :

$$|t^2_{n_2} - x^\lambda_{n_2}| \; / \; |x^2_{n_2} - x^\lambda_{n_2}| \leq \beta ,$$

where :

$$x^\lambda_{n_2} = x^2_{n_2} + \frac{\lambda}{1-\lambda} (x^2_{n_2} - x^2_{n_2-1})$$

(from the lemma, $x^\lambda_{n_2} = \lim_{n\to\infty} x^2_n$)

Analogously, we define (x^3_n) , (x^4_n) , etc...

Now, we consider :

$$(x_n) = (x^0_0, x^0_1, \ldots, x^0_{n_0}, x^1_{n_0+1}, x^1_{n_0+2}, \ldots, x^1_{n_1}, x^1_{n_1+1}, \ldots)$$

Since :

$$\forall\, n \in \mathbf{N} : (x_{n+2} - x_{n+1})/(x_{n+1} - x_n) \in \{\lambda, \mu\} .$$

The sequence (x_n) converges to x , satisfying

$$x \geq x^{\lambda\mu}_{n_0} , \quad x \geq x^{\lambda\mu}_{n_2} , \ldots$$

We obtain :

$$t^0_{n_0} = x^0_{n_0} + x^\lambda_{n_0} - x^0_{n_0} + t^0_{n_0} - x^\lambda_{n_0}$$

$$\geq x^0_{n_0} + (x^\lambda_{n_0} - x^0_{n_0}) - |t^0_{n_0} - x^\lambda_{n_0}|$$

$$\geq x^0_{n_0} + (x^\lambda_{n_0} - x^0_{n_0}) - \beta |x^0_{n_0} - x^\lambda_{n_0}|$$

$$= x^0_{n_0} + (1-\beta)(x^\lambda_{n_0} - x^0_{n_0}) \geq x^0_{n_0} .$$

Similarly : $\quad t^2_{n_2} \geq x^2_{n_2} , \; t^4_{n_4} \geq x^4_{n_4} , \ldots$

The function $\qquad z \to (z - t^0{}_{n_0})/(z - x^0{}_{n_0})$

is increasing; hence :

$$(x - t^0{}_{n_0})/(x - x^0{}_{n_0}) \geq (x^{\lambda\mu}{}_{n_0} - t^0{}_{n_0})/(x^{\lambda\mu}{}_{n_0} - x^0{}_{n_0})$$

Since, for every $n \leq n_0$, $x^0{}_n = x_n$, we have $t_{n_0} = t^0{}_{n_0}$

hence :

$$(x - t_{n_0})/(x - x_{n_0}) \geq (x^{\lambda\mu}{}_{n_0} - t_{n_0})/(x^{\lambda\mu}{}_{n_0} - x_{n_0}) \quad (*)$$

We also obtain :

$$x^{\lambda\mu}{}_{n_0} - t_{n_0} = x^{\lambda\mu}{}_{n_0} - x^\lambda{}_{n_0} + x^\lambda{}_{n_0} - t_{n_0}$$

$$\geq \alpha(x^0{}_{n_0} - x^0{}_{n_0-1})/2 - |x^\lambda{}_{n_0} - t^0{}_{n_0}|$$

$$\geq \alpha(x^0{}_{n_0} - x^0{}_{n_0-1})/2 - \beta|x^\lambda{}_{n_0} - x^\lambda{}_{n_0}|$$

$$= \delta(x^0{}_{n_0} - x^0{}_{n_0-1}) \ .$$

Consequently :

$$(x^{\lambda\mu}{}_{n_0} - t^0{}_{n_0})/(x^{\lambda\mu}{}_{n_0} - x^0{}_{n_0}) \geq \delta(x^0{}_{n_0} - x^0{}_{n_0-1})/(x^{\lambda\mu}{}_{n_0} - x^0{}_{n_0})$$

$$= \frac{\delta(x^0{}_{n_0} - x^0{}_{n_0-1})}{x^\lambda{}_{n_0} - x^0{}_{n_0} - \alpha(x^0{}_{n_0} - x^0{}_{n_0-1})/2} = \Theta$$

With $(*)$, we obtain :

$$(x - t_{n_0})/(x - x_{n_0}) \geq \Theta$$

$$(x - t_{n_2})/(x - x_{n_2}) \geq \Theta$$

Similarly : $\qquad (x - t_{n_4})/(x - x_{n_4}) \geq \Theta$ etc...

Thus A does not accelerate (x_n).

From definition,

$$\forall\, n \in \mathbf{N} \quad (x_{n+2} - x_{n+1})/(x_{n+1} - x_n) \in \{\lambda,\mu\}$$

$$\text{Consequently} \quad (x_n) \in \text{Lin}^{\Delta}{}_{\lambda,\mu}(\mathbf{R}) \ .$$

(ii) Since all the sequences $(x^0{}_n),(x^1{}_n),\dots,(x_n)$ satisfy :

$$\lambda = \frac{x_n - x^{\lambda}{}_n}{x_{n-1} - x_n} \leq \frac{x_n - x}{x_{n-1} - x} \leq \frac{x_n - x^{\mu}{}_n}{x_{n-1} - x^{\mu}{}_n} = \mu$$

where :

$$x = \lim_{n \to \infty} x_n \ , \quad x^{\lambda}{}_n = x_n + \frac{\lambda}{1-\lambda}\,(x_n - x_{n-1}) \ ,$$

$$x^{\mu}{}_n = x_n + \frac{\mu}{1-\mu}\,(x_n - x_{n-1}) \ ,$$

the above proof also holds for $\text{Lim}\lambda,\mu(\mathbf{R})$.

Remarks

1°) In fact, our proof establishes a finer result than theorem 11 :

Let $\lambda,\mu \in \mathbf{R}$, $0 < \lambda < \mu < 1$.

There is no normal transformation which accelerates the convergence of sequences of $\text{Conv}(\mathbf{R})$ such that :

$$\forall\, n \in \mathbf{N} : (x_{n+2} - x_{n+1})/(x_{n+1} - x_n) \in \{\lambda,\mu\}$$

2°) One can easily prove that, if :

$$\forall\, n \in \mathbf{N} : \lambda \leq (x_{n+2} - x_{n+1})/(x_{n+1} - x_n) \leq \mu \quad (A)$$

then :

$$\forall\, n \in \mathbf{N} : \lambda' \leq (x_{n+1} - x)/(x_n - x) \leq \mu' \quad\quad (B)$$

with :

$$\lambda' = (\lambda\mu - \lambda)/(\lambda-1) \ , \quad \mu' = (\lambda\mu-\mu)/(\mu-1) \ .$$

Conversely, if (B) is true with λ',μ' , then (A) is true with :

$$\lambda = (\lambda'\mu'-\lambda')/(\lambda'-1) \ , \quad \mu' = (\lambda'\mu'-\mu')/(\mu'-1) \ .$$

3°) Theorem 11 can be extended in the following form :

Let $\lambda,\mu \in \mathbf{R}$, $0 < \lambda < \mu < 1$, $k \in \mathbf{N}$.

Then there is no k-normal transformation which accelerates $Lin^\Delta_{\lambda\mu}(R)$ or $Lin_{\lambda\mu}(R)$.

$4°$) The proof of theorem 11 does not use remanence. In fact, $Lin_{\lambda,\mu}(R)$ and $Lin^\Delta_{\lambda,\mu}(R)$ are not remanent families. Indeed, if this were the case, no algorithmic transformation could accelerate $Lin_{\lambda,\mu}(R)$ and $Lin^\Delta_{\lambda,\mu}(R)$, but the algorithmic transformation $t_n = x_{n^2}$

accelerates $Lin^\Delta_{\lambda\mu}(R)$ and $Lin_{\lambda,\mu}(R)$ $(\lambda,\mu \in R \quad 0 < \lambda < \mu < 1)$.

$5°$) The family of sequences obtained from $x_{n+1} = f(x_n)$ with f α-Lipschitz $(\alpha < 1)$, contains all the sequences used in our proof. Hence this family is also not accelerable.

Theorem 12

There is no normal transformation which accelerates $SLin(R)$.

Proof

We assume that $N = (f_n)$ is a normal transformation which accelerates $SLin(R)$.

Let (x^0_n) be the sequence defined by :
$$x^0_n = \sum_{i=0}^{n} 1/i!$$

The sequence belongs to $SLin(R)$ and its limit is $x^0 = e$.

From the hypothesis, N accelerates (x^0_n) , thus there exists $n_0 \in N$ such that :
$$|t^0_{n_0} - x^0| \, / \, |x^0_{n_0} - x^0| \leq 1/4 .$$

For every real $x \in]x^0_{n_0} , (x^0_{n_0} + x^0)/2]$, we obtain :
$$|t^0_{n_0} - x| \, / \, |x^0_{n_0} - x| \geq 1/2 .$$

$$(|t^0_{n_0} - x| \geq |x - x^0| - |t^0_{n_0} - x^0| \geq |x^0_{n_0} - x^0|/2 - |x^0_{n_0} - x^0|/4$$

$$= |x^0_{n_0} - x^0|/4 \geq |x^0_{n_0} - x|/2)$$

We define the sequence (x^1_n) by :

$x^1_n = x^0_n$ for every $n \leq n_0$,

$x^1_n = x^1_{n-1} + 1/2n!$ for every $n > n_0$.

The sequence $(x^1_n) \in$ **SLin(R)** and its limit is :

$$x^1 = \frac{x^0 + x^0_{n_0}}{2}$$

From the hypothesis, N accelerates (x^1_n) , thus there exists $n_1 > n_0$ such that :

$$|t^1_{n_1} - x^1| \; / \; |x^1_{n_1} - x^1| \leq 1/4 \; ;$$

For every real $x \in \;]x^1_{n_1} \; , \; (x^1_{n_1} + x^1)/2]$, we obtain :

$$|t^1_{n_1} - x| \; / \; |x^1_{n_1} - x| \geq 1/2 \; .$$

Similarly, we define (x^2_n) , (x^3_n) , ...

Let (x_n) be the following sequence :

$$(x_n) = (x^0_0, x^0_1, \ldots, x^0_{n_0}, x^1_{n_0+1}, \ldots, x^1_{n_1}, x^2_{n_1+1}, \ldots)$$

This sequence is increasing and converges to x such that :

$$\forall \; i \in \mathbf{N} : x \in \;]x^i_{n_i} \; , \; (x^i_{n_i} + x^i)/2].$$

This sequence belongs to **SLin(R)**, but :

$$\forall \; i \in \mathbf{N} : |t_{n_i} - x| \; / \; |x_{n_i} - x| \geq 1/2 \; .$$

Consequently, (t_n) does not accelerate (x_n) .

Remark

Contrarily to $Lin_{\lambda\mu}(R)$, $Lin^\Delta_{\lambda\mu}(R)$ the family **SLin(R)** is accelerable with k-normal transformations $(k \leq 1)$ (this is the first case of this sort). Indeed, the transformation $t_n = x_{n+1}$ accelerates **SLin(R)**.

5 - FAMILIES OF LOGARITHMICALLY CONVERGENT SEQUENCES

In [23] D.A. SMITH and W.F. FORD give a detailed numerical study of the methods for accelerating logarithmic convergence. They say that, based on empirical evidence, good test of applicability to logarithmic convergence is exactness on series of Cordellier's types (F2) and (F3) ([6]), and they ask : "What the proper analog of Germain-Bonne's theorem ([17]) for logarithmic convergence ? In particular, does exactness on series of types (F2) and (F3) imply acceleration for all logarithmic series ?".

The method of remanence gives the answer : NO, there is no algorithm which accelerates all logarithmic series or sequences (theorem 13); there is no proper analog of Germain-Bonne's theorem; the empirical test of applicability to logarithmic convergence is not always good.

The conclusion of this section is that logarithmically convergent sequences are difficult to accelerate : it is not possible to accelerate all of them with only one transformation.

The various results given here were first published in [16], Theorem 13 is a generalization of a result of [14]. The work of C. KOWALEWSKI ([20],[21]) gives more detail on the problem.

Definition

We set :

$$\text{Log} = \{(x_n) \in \text{Conv}^*(R) \mid \lim_{n \to \infty} \frac{x_{n+1} - x}{x_n - x} = 1\}$$

and

$$\text{LogSF} = \{(x_n) \in \text{Conv}^*(R) \mid \lim_{n \to \infty} \frac{x_{n+1} - x}{x_n - x} = \lim_{n \to \infty} \frac{x_{n+2} - x_{n+1}}{x_{n+1} - x_n} = 1\}$$

Theorem 13

The family Log SF satisfies the generalised remanence, hence there is no algorithmic transformation which accelerates LogSF or Log.

Lemma

Let $p_0 \in N$, a,b,c,r $\in R$ be such that : a < b < c ; 0 < r < 1 ;

$$(c-b)/(1-r) < (c-a) .$$

There exists a sequence (x_n) such that :

(i) $x_{p_0} = c$, $x_{p_0 + 1} = b$;

(ii) $(x_n) \to a$;

(iii) $\forall\ n \geq p_0 : r \leq (x_{n+2} - x_{n+1})/(x_{n+1} - x_n) \leq 1$;

(iv) $(x_{n+2} - x_{n+1})/(x_{n+1} - x_n) \to 1$;

(v) $\forall\ n \geq p_0 : r \leq (x_{n+1} - a)/(x_n - a) \leq 1$;

(vi) $(x_{n+1} - a)/(x_n - a) \to 1$.

Proof of the lemma

Without loss of generality, one can assume that $p_0 = 0$.

Let (r_n) be a real, increasing sequence which converges to 1 and such that $r_0 = r$.

We shall construct convergent sequences $(x^0{}_n), (x^1{}_n), \ldots, (x^i{}_n) \ldots$ (with limits $x^0, x^1, \ldots, x^i, \ldots$ respectively), and then define the desired sequence (x_n).

Construction of $(x^0{}_n)$

We put $s_0 = r$, $x^0{}_0 = c$ and, for every $n \geq 1$,

$$x^0{}_n = c - (c-b)(1 + s_0 + s^2{}_0 + \ldots + s^{n-1}{}_0).$$

We have :

$$x^0 = c - \frac{c-b}{1-s_0} = c - \frac{c-b}{1-r} > c - (c-a) = a$$

Let $m_{-1} = 0$.

Construction of $(x^{i+1}{}_n)$

There exists an integer $m_i \geq m_{i-1}$ and a real s_{i+1} such that :

(a_{i+1}) $r_{i+1} \leq s_{i+1} \leq 1$

(b_{i+1}) $|x^i{}_{m_i} - x^i| \leq 1/2^i$

(c_{i+1}) $x^i{}_{m_i - 1} - \dfrac{x^i{}_{m_i - 1} - x^i{}_{m_i}}{1 - s_{i+1}} = \dfrac{x^i + a}{2}$.

First, we determine m_i such that :

$$|x^i_{m_i} - x^i| \leq \frac{1}{2^i} \quad , \quad x^i_{m_i-1} - \frac{x^i_{m_i} - x^i_{m_i}}{1 - r_{i+1}} \geq \frac{x^i + a}{2}$$

Then, we choose s_{i+1}.

The sequence (x^{i+1}_n) is defined by :

$$x^{i+1}_n = \begin{cases} x^i_n \quad \text{for} \quad n \in \{0,1,\ldots,m_i\} \\ x^i_{m_i-1} - (x^i_{m_i-1} - x^i_{m_i})(1 + s_{i+1} + s^2_{i+1} + \ldots + s^{n-m_i}_{i+1}) \\ \qquad\qquad\qquad \text{for} \quad n > m_i \end{cases}$$

We have :

$$x^{i+1} = x^i_{m_i-1} - (x^i_{m_i-1} - x^i_{m_i})/(1 - s_{i+1}) = (x^i + a)/2 > a .$$

When all the sequences $(x^0_n),(x^1_n),\ldots,(x^i_n),\ldots$ are defined, we put

$$(x_n) = (x^0_0 , x^0_1,\ldots, x^0_{m_0} , x^1_{m_0-1} ,\ldots, x^1_{m_1} , x^2_{m_1+1},\ldots).$$

Now we proove that, for this sequence (i), (ii), (iii), (iv), (v), (vi) are true.

(i)   By construction, $x^0_0 = c$, $x^0_1 = b$. Since $p_0 = 0$, (i) is true.

(ii)   The sequence (x_n) decreases.
Also, from $(b_1),(b_2),\ldots,(b_i),\ldots$ we obtain :

$$\forall i \geq 1 : |x^i - x_{m_i}| \leq 1/2^i ;$$

While by construction, we obtain :

$$\forall i \geq 1 : x^i = \frac{x^{i-1} + a}{2}$$

Therefore (ii) is true.

(iii)   and   (iv)

If $m \in \{m_i-1, m_i, \ldots, m_{i+1} - 2\}$,   then

$$(x_{m+2} - x_{m+1})/(x_{m+1} - x_m) = s_{i+1} .$$

From $(a_1), (a_2), \ldots, (a_i), \ldots,$ we obtain :

$$\forall \; i \in \mathbf{N} , \; r_i \leq s_i \leq 1 .$$

Consequently, (iii) and (iv) are true.

(v)   and   (vi)

If $m \in \{m_i, m_i+1, \ldots, m_{i+1}-1\}$

then $(x_{m+1} - x^{i+1})/(x_m - x^{i+1}) = s_{i+1} .$

Since $a < x^{i+1} < x_{m+1} < x_m$,   we obtain :

$$s_{i+1} \leq (x_{m+1} - a)/(x_m - a) \leq 1 ,$$

and thus (v) and (vi) are true.

Proof of the theorem

In   (GR(a))   we take $(x^n) = (1/(n+1))$.

Let (v_n) be an increasing sequence of real numbers which converges to 1 .

$(GR(a)(0^0))$

We define $(x^0{}_n)$ by $(x^0{}_n) = (1 + 1/(n+1))$.

$(GR(a)(1^0))$

Let $m_0 \geq 0$. There exists $p_0 \geq m_0$ such that :

(A1) $(x^0{}_{p_0} - x^0{}_{p_0+1})/(x^0{}_{p_0} - x^1) < 1 - v_0$;

(B1) $|x^0{}_{p_0} - x^0| \leq 1/2^0$.

Let $(y^1{}_n)$ be the sequence given by the lemma with p_0 ,

$$a = x^1 , \; b = x^0{}_{p_0+1} , \; c = x^0{}_{p_0} , \; r = v_0$$

(this sequence exists since (A1) is satisfied).

The sequence $(x^1{}_n)$ is then defined by :

$$x^1{}_n = x^0{}_n \quad \text{if} \quad n \in \{0,1,\ldots,p_0\} \; ,$$

$$x^1{}_n = y^1{}_n \quad \text{if} \quad n > p_0.$$

...

$(RG(a)(i^0))$

Let $m_{i-1} > p_{i-2}$. There exists $p_{i-1} > m_{i-1}$ such that :

(A_i) $\qquad (x^{i-1}{}_{p_{i-1}} - x^{i-1}{}_{p_{i-1}+1}) \, / \, (x^{i-1}{}_{p_{i-1}} - x^i) < 1 - v_{i-1}$,

(B_i) $\qquad |x^{i-1}{}_{p_{p-1}} - x^{i-1}| \leq 1/2^{i-1}$

Let $(y^i{}_n)$ be the sequence given by the lemme with p_{i-1} ,

$$a = x^i \; , \; b = x^{i-1}{}_{p_{i-1}+1} \; , \; c = x^{i-1}{}_{p_{i-1}} \; , \; r = v_{i-1}$$

The sequence $(x^i{}_n)$ is then defined by :

$$x^i{}_n = x^{i-1}{}_n \quad \text{if} \quad n \in \{0,1,\ldots,p_{i-1}\}$$

$$x^i{}_n = y^i{}_n \quad \text{if} \quad n > p_{i-1} \; .$$

$(RG(b))$

Now we must prove that the sequence :

$$(x_n) = (x^0{}_0, x^0{}_1, \ldots, x^0{}_{p_0}, x^1{}_{p_0-1}, \ldots, x^1{}_{p_1}, x^2{}_{p_1+1}, \ldots)$$

is a sequence of **LogSF**.

From $(B1),(B2),\ldots,(Bi),\ldots,$ we obtain that (x_n) converge to 0 .

From $(A1),(A2),\ldots,(Ai),\ldots,$ and the properties (i) and (iii) of the lemma, we obtain :

$$\lim_{n \to \infty} \frac{x_{n+2} - x_{n+1}}{x_{n+1} - x_n} = 1$$

As in the proof of (v) and (vi) of the lemma, we obtain :

$$\lim_{n \to \infty} \frac{x_{n+1} - x}{x_n - x} = 1$$

Remark

LogSF is not accelerable with one sequence transformation. However, many sequences are accelerated by specific transformation ([1],[3],[5],[6],[23] and [25]) . Now the problem is no longer how to accelerate LogSF, but what are the accelerable subsets of LogSF (and by what transformation ?).

The work of C. KOWALEWSKI is a contribution to the answer of the problem ([20],[21]).

6 - TABLE OF RESULTS

When E = R , none of the following families is accelerable :

	Definition	result	
Conv	p. 154	Th. 4	p. 161
Conv*	p. 154	Th. 5	p. 162
Mon$^-$, Mon$^+$, Mon	p. 164	Th. 6	p. 165
Mon^{-*}, Mon^{+*}, Mon*	p. 164	Th. 7	p. 165
Mon^-_k, Mon^+_k, Mon	p. 164	Th. 8	p. 166
Mon$^{-*}_k$, Mon$^{+*}_k$, Mon$^*_k$	p. 164	Th. 8	p. 166
Alt A , Alt B , Osc A , Osc B	p. 170	Th. 9,10	p. 171
Lin$^\Delta_{\lambda,\mu}$, Lin$_{\lambda\mu}$, Lin$_{\lambda\mu}$ ($0 \leq \lambda < \mu \leq 1$)	p. 170	Th. 11	p. 174
SLin	p. 174	Th. 12	p. 179
Log , LogSF	p. 181	Th. 13	p. 181

CONCLUSION

For every type of sequence families, the acceleration results have to determine as precisely as possible where the frontier between acce-lerability and non accelerability is.

Positive results try to obtain bigger accelerable families, negative results try to obtain smaller non accelerable families.

This two types of results never meet (since we know there is no maximal accelerable families, nor minimal non—accelerable families) but about families of monotone sequences, of alternating sequences, of linearly or logarithmically convergent sequences we have a fairly good idea of the frontier.

REFERENCES

[1]   BREZINSKI C.
      "Accélération des suites à convergence logarithmique"
      C.R. Acad. Sc. Paris, A, 273, 1971, pp. 772-774.

[2]   BREZINSKI C.
      "Génération de suites totalement monotones et oscillantes"
      C.R. Acad. Sc. Paris, A, 280, 1975, pp. 729-731.

[3]   BREZINSKI C.
      "Accélération de la convergence en analyse numérique"
      Lecture Notes in Mathematics, 584, Springer-Verlag,
      Heidelberg, 1977.

[4]   BREZINSKI C.
      "Convergence Acceleration of some sequences by the
      ε-algorithm"
      Num. Math. 29, 1978, pp. 173-177.

[5]   BREZINSKI C.
      "Algorithmes d'accélération de la convergence, étude numérique"
      Technip, Paris, 1978.

[6]   CORDELLIER F.
      "Caractérisation des suites que la première étape du
      θ-algorithme transforme en suites constantes"
      C.R. Acad. Sc. Paris, A, 284, 1977, pp. 389-392.

[7]   DELAHAYE J.P.
      "Liens entre la suite du rapport des erreurs et celle
      du rapport des différences : démonstrations"
      Publication ANO n° 14, Univ. des Sc. et Techn. de Lille.

[8]   DELAHAYE J.P.
      "Liens entre la suite du rapport des erreurs et celle
      du rapport des différences"
      C.R. Acad. Sc. Paris, A, 290, 1980, pp. 343-346

[9]   DELAHAYE J.P.
      "Accélération des suites dont le rapport des erreurs est
      borné"
      Calcolo, 18, 1981, pp. 103-116

[10] DELAHAYE J.P.
      "Optimalité du procédé Δ^2 d'Aitken pour l'accélération de
      la convergence linéaire"
      RAIRO An. Num. 15, 1981, pp. 321-330.

[11] DELAHAYE J.P.
      "Les divers types de transformations algorithmiques de
      suites"
      Publication ANO, n° 69, Univ. Sc. et Techn. de Lille, 1982.

[12]   DELAHAYE J.P.
"Familles de suites alternées et propriétés de rémanence"
Colloque d'Analyse Numérique de Belgodère, mai 1982.

[13]   DELAHAYE J.P. et GERMAIN-BONNE B.
"Quelques résultats négatifs en accélération de la convergence"
Publication ANO n° 6, Univ. Sc. et Techn. de Lille, 1979.

[14]   DELAHAYE J.P. et GERMAIN-BONNE B.
"Résultats négatifs concernant les algorithmes d'accélération
de la convergence"
Publication n° 337 du Labo. IMAG de l'Univ. Sc. et Méd. de
Grenoble, 1980.

[15]   DELAHAYE J.P. et GERMAIN-BONNE B.
Résultats négatifs en accélération de la convergence"
Num. Math., 35, 1980, pp. 443-457.

[16]   DELAHAYE J.P. et GERMAIN-BONNE B.
"The set of logarithmically convergent sequences cannot
be accelerated"
SIAM, Num. An. 19, 1982, pp. 840-844.

[17]   GERMAIN-BONNE B.
"Estimation de la limite des suites et formalisation des
procédés d'accélération de convergence"
Thèse, Lille, 19678.

[18]   GILEWICZ J.
"Approximants de Padé"
Lecture Notes in Mathematics, 667, Springer-Verlag,
Heidelberg, 1978.

[19]   HILLION P.
"Méthode d'Aitken itérée pour suites oscillantes
d'approximations"
C.R. Acad. Sc. Paris, A, 280, 1975, pp. 1701-1703.

[20]   KOWALEWSKI C.
"Possibilités d'accélération de la convergence logarithmique"
Thèse de 3° cycle, Lille, 1981.

[21]   KOWALEWSKI C.
"Accélération de la convergence pour certaines suites à
convergence logarithmique"
in Padé Approximation and its applications,
Amsterdam, 1980, Lectures Notes in Mathematics, 888,
Springer-Verlag, Heidelberg, 1981, pp. 263-272.

[22]   PENNACHI R.
"Le transformazioni rationali di una successionne"
Calcolo, 5, 1968, pp. 37,50.

[23]   SMITH D.A. and FORD W.F.
       "Acceleration of linear and logarithmic convergence"
       S.I.A.M., Numer. An., 16, 1979, pp. 223-240.

[24]   TROJAN J.M.
       "An upper bound on the acceleration of convergence"
       First French-Polish Meeting on Padé Approximation and
       convergence acceleration techniques, Varsovie, Pologne,
       1-4 Juin 1981.

[25]   WIMP J.
       "Sequence transformations and their applications"
       Academic Press, New York, 1981

Chapter 6

**Accelerating The Convergence
of Linear Sequences**

INTRODUCTION

In this chapter, we study some problems related to accelerating the convergence of families of linearly convergent sequences.

The results of § 1 give a characterisation of linearly convergent sequences and of periodico-linearly convergent sequences. These characterisations allow us (in section 2), to use the algorithms of chapter 4 in order to obtain some new methods of accelerating the convergence, well adapted to periodico-linearly convergent sequences.

The difficulties in § 2 lead us to think that when we enlarge the family of all linearly convergent sequences, we obtain a non-accelerable family. Indeed, in § 3, we show that the Δ^2 of Aitken is optimal in at least three different meanings.

1.- It is the algebraically simplest transformation which accelerates the convergence of linear sequences.

2.- Its domain of efficiency cannot be enlarged.

3.- The degree of acceleration obtained with it is the best possible.

The results of § 1 has been published in [11] and [4], the results of § 2 in [10] and [15], the results of § 3 in [16].

1 - LINEARLY CONVERGENT AND PERIODICO-LINEARLY CONVERGENT SEQUENCES

Let (x_n) be a given real or complex sequence. We consider $(x_{n+1}-x)/(x_n-x)$ (x the limit of (x_n)) and $(x_{n+2}-x_{n+1})/(x_{n+1}-x_n)$.

We study their relative behaviors concerning the convergence (a) , then concerning problems of pseudo-periodicity (b).

The results show that in almost all cases these two sequences have similar behaviors. The results allow to transform some statements in the acceleration of convergence with hypothesis that are hard to verify into equivalent statements with hypothesis which are easy to verify. Examples of such statement transformations are given in (c).

(a) Problems on convergence

Let (x_n) be a real (or complex) sequence such that :

$$\forall \; n \in \mathbf{N} \qquad x_{n+1} \neq x_n$$

We shall prove the following theorem.

Theorem 1

Let $\lambda \in \mathbf{C}$, $|\lambda| \neq 1$. Then :

| there exists $x \in \mathbf{C}$ such that $$\left(\frac{x_{n+1}-x}{x_n-x}\right)$$ converges and $$\lim_{n\to\infty} \frac{x_{n+1}-x}{x_n-x} = \lambda$$ | $\Longleftrightarrow$ | $$\left(\frac{x_{n+2}-x_{n+1}}{x_{n+1}-x_n}\right)$$ converges and $$\lim_{n\to\infty} \frac{x_{n+2}-x_{n+1}}{x_{n+1}-x_n} = \lambda$$ |

Remarks

1°) If $|\lambda| = 1$, then the theorem is not true. Indeed :

    a) counterexample for $\Longleftarrow$

$$x_{2n} = \frac{1}{n^2} \quad , \quad x_{2n+1} = \frac{-1}{n}$$

    b) counterexample for $\Longrightarrow$

$$x_{2n} = \frac{1}{3n} \quad , \quad x_{2n+1} = \frac{1}{3n+1}$$

The theorem is a consequence of propositions 1, 2 and 3, which give some additional information (for example, that if $|\lambda| > 1$, then in the implication " $\Longleftarrow$ " , "there exists $x \in \mathbf{C}$" may be replaced by "for every $x \in \mathbf{C}$")

2°) Another proof of theorem 1 is given in $[30]$.

Proposition 1 ([20])

If there exists $x \in \mathbf{C}$ and $\lambda \in \mathbf{C}$, $\lambda \neq 1$, such that

$$\frac{x_{n+1}-x}{x_n-x} \quad \text{converges and} \quad \lim_{n\to\infty} \frac{x_{n+1}-x}{x_n-x} = \lambda , \quad \text{then}$$

$$\frac{x_{n+2}-x_{n+1}}{x_{n+1}-x_n} \quad \text{converges and} \quad \lim_{n\to\infty} \frac{x_{n+2}-x_{n+1}}{x_{n+1}-x_n} = \lambda .$$

Proof

$$\frac{x_{n+2}-x_{n+1}}{x_{n+1}-x_n} = \frac{x_{n+1}-x}{x_n-x} \; \frac{\dfrac{x_{n+2}-x}{x_{n+1}-x} - 1}{\dfrac{x_{n+1}-x}{x_n-x} - 1} \xrightarrow[n\to\infty]{} \lambda \; \frac{\lambda-1}{\lambda-1} = \lambda$$

Remark

This proof is also valuable when (x_n) is any sequence in a topological field. Consequently, if $\mathbf{K}$ is a given field with the discrete
topology, we obtain that :

$$\left[\exists \; \lambda \in \mathbf{K} ; \; \lambda \neq 1 ; \; \exists \; x \in \mathbf{K} \; \exists \, N \in \mathbf{N} , \right.$$

$$\left. \forall \, n \geq N , \; \frac{x_{n+1}-x}{x_n-x} = \lambda \; \right] \Rightarrow$$

$$\left[\exists \, N \in \mathbf{N} , \; \forall \, n \geq N \; \frac{x_{n+2}-x_{n+1}}{x_{n+1}-x_n} = \lambda \; \right]$$

For every $n \in \mathbf{N}$, we set :

$$\lambda_n = \frac{x_{n+2}-x_{n+1}}{x_{n+1}-x_n}$$

For every $n \in \mathbf{N}$, $k \in \mathbf{N}$, we set :

$$\mu^k_n = \lambda_n + \lambda_n \; \lambda_{n+1} + \dots + \lambda_n \; \lambda_{n+1} \dots \lambda_{n+k} = \sum_{j=0}^{k} \prod_{i=0}^{j} \lambda_{n+i}$$

Lemma 1

a) For every $n \in \mathbf{N}$, $k \in \mathbf{N}$, $x_{n+k+2} = x_{n+1} + \mu^k{}_n(x_{n+1}-x_n)$

b) If for   m   fixed,   the sequence $(\mu^k{}_m)_{k \in \mathbf{N}}$ converges to μ_m ,
    then (x_n) converges to x such that :

$$x = x_{m+1} + \mu_m(x_{m+1}-x_m)$$

and :

$$\mu_m \neq -1 \Rightarrow \frac{x_{m+1}-x}{x_m-x} = \frac{\mu_m}{1+\mu_m} \; .$$

Proof

We have : $x_{n+2} - x_{n+1} = \lambda_n(x_{n+1}-x_n)$;

$$x_{n+3} - x_{n+2} = \lambda_{n+1} \lambda_n(x_{n+1}-x_n) \; ;$$

$$\cdots \quad \cdots \quad \cdots$$

$$x_{n+k+2} - x_{n+k+1} = \lambda_{n+k} \lambda_{n+k-1} \cdots \lambda_n(x_{n+1}-x_n)$$

Thus :

$$x_{n+k+2} = x_{n+1} + \mu^k{}_n(x_{n+1}-x_n)$$

Part b then follows immediatly.

Remark

The proof and thus lemma 1 are still true without any modification for
sequences in a topological field.

Lemma 2

Let (λ_n) be a real or complex convergent sequences with limit λ
$|\lambda| < 1$. We set $\mu^k{}_n = \lambda_n + \lambda_n \lambda_{n+1} + \cdots + \lambda_n \lambda_{n+1} \cdots \lambda_{n+k}$. Then,
for every $n \in \mathbf{N}$,   the sequence $(\mu^k{}_n)_{k \in \mathbf{N}}$ converges (if $\mu^k{}_n$ is
viewed as a serie, the convergence is absolute).

Furthermore, the sequence (μ_n) defined by

$$\mu_n = \lim_{k \to \infty} \mu^k{}_n$$

converges to $\lambda/(1-\lambda)$.

Proof

The absolute convergence is obvious. We will show the other statement
Let $\varepsilon \in \mathbf{R}^{+*}$.

Let $\lambda' \in \mathbf{R}$ be such that $|\lambda| < \lambda' < 1$.

Let $n_0 \in \mathbf{N}$ be such that $n \geq n_0 \Rightarrow |\lambda_n| \leq \lambda'$.

Let $k_0 \in \mathbf{N}$ be such that :

i) $\quad \left| \dfrac{\lambda^{k_0+1}}{\lambda-1} \right| \leq \varepsilon/3$

ii) $\forall\, n \geq n_0 \quad \forall\, k \geq k_0$

$\quad |\lambda_n \lambda_{n+1} \cdots \lambda_{n+k_0} + \cdots + \lambda_n \lambda_{n+1} \cdots \lambda_{n+k}| \leq \varepsilon/3$.

The integer k_0 exists, since :

$\quad \forall\, n \geq n_0$

$|\lambda_n \lambda_{n+1} \cdots \lambda_{n+k_0} + \cdots + \lambda_n \lambda_{n+1} \cdots \lambda_{n+k}|$

$$\leq \lambda'^{k_0+1} + \lambda'^{k_0+2} + \cdots \leq \frac{\lambda'^{k_0+1}}{1-\lambda'}$$

Let $n_1 \geq n_0$ be such that for every $n \geq n_y$

$|(\lambda_n-\lambda) + (\lambda_n \lambda_{n+1}-\lambda^2) + \cdots + (\lambda_n \lambda_{n+1} \cdots \lambda_{n+k_0-1} - \lambda^{k_0})| \leq \varepsilon/3$

(n_1 exists, since the sequence (λ_n) converges to λ , the sequence $\lambda_n \lambda_{n+1}$ converges to $\lambda^2, \ldots,$ the sequence $\lambda_n \lambda_{n+1} \cdots \lambda_{n+k_0-1}$ converges to λ^{k_0}).

For every $n \geq n_1$ and every $k \geq k_0$:

$\quad |\lambda_n + \lambda_n \lambda_{n+1} + \cdots + \lambda_n \lambda_{n+1} \cdots \lambda_{n+k} - \lambda/(1-\lambda)|$

$= |(\lambda_n - \lambda) + (\lambda_n \lambda_{n+1} - \lambda^2) + \cdots + (\lambda_n \lambda_{n+1} \cdots \lambda_{n+k_0-1} - \lambda^{k_0})$

$\qquad\qquad\qquad\qquad - \lambda^{k_0+1}/(1-\lambda)$

$+ \lambda_n \lambda_{n+1} \cdots \lambda_{n+k_0} + \cdots + \lambda_n \lambda_{n+1} \cdots \lambda_{n+k}| \leq \varepsilon/3 + \varepsilon/3 + \varepsilon/3 = \varepsilon.$

Remark

This proof and thus the lemma 2 are still true without any modification for sequences in a valued field.

Proposition 2

Let $\lambda \in \mathbf{C}$, $|\lambda| < 1$. If :

$$\frac{x_{n+2} - x_{n+1}}{x_{n+1} - x_n} \quad \text{converges and if} \quad \lambda = \lim_{n \to \infty} \frac{x_{n+2} - x_{n+1}}{x_{n+1} - x_n} ,$$

then :

the sequence (x_n) converges to a limit x such that :

$$\frac{x_{n+1} - x}{x_n - x} \quad \text{converges and} \quad \lambda = \lim_{n \to \infty} \frac{x_{n+1} - x}{x_n - x} .$$

Proof

We use the notation given just before lemma 1.   From lemma 2, we obtain that, for every $n \in \mathbf{N}$:

* $(\mu^k{}_n)_{k \in \mathbf{N}}$ is convergent to a limit μ_n ;

*   there exists n_0 such that : $n \geq n_0 \Rightarrow \mu_n \neq -1$.

Consequently, with lemma 1 and lemma 2, we obtain :

$$\lim_{n \to \infty} \frac{x_{n+1} - x}{x_n - x} = \lim_{n \to \infty} \frac{\mu_n}{1 + \mu_n} = \lambda$$

Remarks

1°)   As in lemmas 1 and 2, the result is true for sequences in a valued field.

2°)   When $o < \lambda < 1$, proposition 2 is a consequence of [7]   (see also [4]).

3°)   With real sequences, proposition 2 is used (without proof) in [27].

Lemma 3

Let (α_n) be a real or complex convergent sequence with limit α, $|\alpha| < 1$. Then the sequence defined by :

$$y_n = \alpha_n + \alpha_n\,\alpha_{n-1} + \ldots + \alpha_n\,\alpha_{n-1}\ldots\alpha_1 \quad \text{converges to} \quad \alpha/(1-\alpha).$$

Proof

Let $\varepsilon \in \mathbf{R}^{+*}$.

Let $\alpha' \in \mathbf{R}$ be such that : $|\alpha| < \alpha' < 1$.

There exists n_0 such that $n \geq n_0 \Rightarrow |\alpha_n| < \alpha'$.

Let $k_0 \geq n_0$ be such that :

i)
$$\left|\frac{\alpha^{k_0}}{\alpha - 1}\right| \leq \varepsilon/3$$

ii) $\forall\, n \geq k_0$ $\left|\alpha_n\,\alpha_{n-1}\ldots\alpha_{n-k_0}+1 + \ldots + \alpha_n\,\alpha_{n-1}\ldots\alpha_1\right| \leq \varepsilon/3$

The integer k_0 exists, since :

$\forall\, n \geq n_0$ $\left|\alpha_n\,\alpha_{n-1}\ldots\alpha_{n-k}+_1 + \ldots + \alpha_n\,\alpha_{n-1}\ldots\alpha_1\right|$

$\leq \left|\alpha_n\,\alpha_{n-1}\ldots\alpha_{n-k_0}+1 + \ldots + \alpha_n\,\alpha_{n-1}\ldots\alpha_{n_0}+1\right| + \left|\alpha_n\,\alpha_{n-1}\ldots\alpha_{n_0}\right|$

$\left|1 + \alpha_{n_0}-1 + \ldots + \alpha_{n_0}-1\ldots\alpha_1\right|$

$\leq \alpha'^{k_0}(1 + \alpha' + \ldots + \alpha'^{n-n_0-k_0}) + \alpha'^{n-n_0+1}$

$$\left|1 + \alpha_{n_0}-1 + \ldots + \alpha_{n_0}-1\ldots\alpha_1\right|$$

$\leq \alpha'^{k_0}/(1-\alpha') + \alpha'^{n-n_0+1}\left|1 + \alpha_{n_0}-1 + \ldots + \alpha_{n_0}-1\ldots\alpha_1\right|$

Let $n_1 \geq k_0$ be such that, for every $n \geq n_1$:

$$\left|(\alpha_n-\alpha) + (\alpha_n\,\alpha_{n-1}-\alpha^2) + \ldots + (\alpha_n\,\alpha_{n-1}\ldots\alpha_{n-k_0}+2 - \alpha^{k_0-1})\right| \leq \varepsilon/3.$$

The integer n_1 exists, since the sequence

$$\alpha_n + \alpha_n\,\alpha_{n-1} + \ldots + \alpha_n\,\alpha_{n-1}\ldots\alpha_{n-k_0}+2$$

converges to $\alpha + \alpha^2 + \ldots + \alpha^{k_0-1}$

For every $n \geq n_1$, we obtain that :

$$\left| \alpha_n + \alpha_n \alpha_{n-1} + \ldots + \alpha_n \alpha_{n-1} \ldots \alpha_1 - \frac{\alpha}{1-\alpha} \right|$$

$$= \left| (\alpha_n - \alpha) + (\alpha_n \alpha_{n-1} - \alpha^2) + \ldots + (\alpha_n \alpha_{n-1} \ldots \alpha_{n-k_0+2} - \alpha^{k_0-1}) \right.$$

$$+ (\alpha_n \alpha_{n+1} \ldots \alpha_{n-k_0+1}) + \ldots + (\alpha_n \alpha_{n-1} \ldots \alpha_1) - \left. \frac{\alpha^{k_0}}{1-\alpha} \right|$$

$$\leq \varepsilon/3 + \varepsilon/3 + \varepsilon/3 = \varepsilon.$$

Remark

This proof and thus lemma 3 are still true without any modification for sequences in a valued field.

Proposition 3

Let $\lambda \in \mathbf{C}$, $|\lambda| > 1$.   If :

$$\frac{x_{n+2} - x_{n+1}}{x_{n+1} - x_n} \quad \text{converges and if} \quad \lambda = \lim_{n \to \infty} \frac{x_{n+2} - x_{n+1}}{x_{n+1} - x_n} ,$$

Then :

the sequence (x_n) satisfies : $\lim_{n \to \infty} |x_n| = + \infty$,   and
for every $x \in \mathbf{C}$:

$$\frac{x_{n+1} - x}{x_n - x} \quad \text{converges to} \quad \lambda$$

Proof

We use the notation of lemma 1 :

$$\frac{x_{n+2} - x_1}{x_{n+1} - x_1} = \frac{(\lambda_0 + \lambda_0 \lambda_1 + \ldots + \lambda_0 \lambda_1 \ldots \lambda_n)(x_1 - x_0)}{(\lambda_0 + \lambda_0 \lambda_1 + \ldots + \lambda_0 \lambda_1 \ldots \lambda_{n-1})(x_1 - x_0)}$$

We set $\alpha_n = 1/\lambda_n$. We obtain $\lim\limits_{n\to\infty} \alpha_n = \alpha = \dfrac{1}{\lambda}$ and :

$$\frac{x_{n+2} - x_1}{x_{n+1} - x_1} = \frac{1 + \alpha_n + \alpha_n\,\alpha_{n-1} + \ldots + \alpha_n\,\alpha_{n-1}\,\cdots\,\alpha_1}{\alpha_n + \alpha_n\,\alpha_{n-1} + \ldots + \alpha_n\,\alpha_{n-1}\,\cdots\,\alpha_1} \;\cdot$$

From lemma 3 , we obtain :

$$\lim_{n\to\infty} \frac{x_{n+2} - x_1}{x_{n+1} - x_1} = \frac{1 + \dfrac{\alpha}{1-\alpha}}{\dfrac{\alpha}{1-\alpha}} = \frac{1}{\alpha} = \lambda$$

We have :

$$\frac{x_{n+2} - x_1}{(x_1 - x_0)(\lambda_0\,\lambda_1\ldots\lambda_n)} = \frac{\lambda_0 + \lambda_0\,\lambda_1 + \ldots + \lambda_0\,\lambda_1\,\cdots\,\lambda_n}{\lambda_0\,\lambda_1\,\cdots\,\lambda_n} =$$

$$= 1 + \alpha_n + \alpha_n\,\alpha_{n-1} + \ldots + \alpha_n\,\alpha_{n-1}\,\cdots\,\alpha_1 \;\cdot$$

Hence :
$$\lim_{n\to\infty} \frac{x_{n+2} - x_1}{(x_1 - x_0)(\lambda_0\,\lambda_1\,\cdots\,\lambda_n)} = \frac{1}{1-\alpha} \neq 0 \;.$$

Thus (with $\lim\limits_{n\to\infty} |\lambda_0\,\lambda_1\,\cdots\,\lambda_n| = +\infty$) , we obtain that :

$$\lim_{n\to\infty} |x_{n+2} - x_1| = +\infty \;;$$

Then, we have :

$$\lim_{n\to\infty} |x_{n+2}| = +\infty \quad \text{and} \quad \forall\, x \in \mathbf{C} \;\; \lim_{n\to\infty} \frac{x_{n+1} - x}{x_n - x} = \lim_{n\to\infty} \frac{x_{n+1}}{x_n} = \lambda \;.$$

Remark

The result is still true for sequences in a valued field.

(b) Problems of pseudo-periodicity

In this section, $k \in \mathbf{N}^*$ and (x_n) is always a real or complex sequence such that :

$$\forall\, n \in \mathbf{N} : x_{n+1} - x_n \neq 0.$$

Theorem 2

Let $(\lambda^0, \lambda^1, \ldots, \lambda^{k-1}) \in \mathbb{C}^k$ and $(\beta^0, \beta^1, \ldots, \beta^{k-1}) \in \mathbb{C}^k$ be such that :

(B) $\beta^0 \neq 1$, $\beta^1 \neq 1$, $\ldots$, $\beta^{k-1} \neq 1$

(L)

$$\left| \begin{array}{l} |\lambda^0\ \lambda^1\ \ldots\ \lambda^{k-1}| \neq 1 \ ; \\[4pt] 1 + \lambda^0 + \lambda^0\ \lambda^1 + \ldots + \lambda^0\ \lambda^1\ \ldots\ \lambda^{k-2} \neq 0 . \\[4pt] 1 + \lambda^1 + \lambda^1\ \lambda^2 + \ldots + \lambda^1\ \lambda^2\ \ldots\ \lambda^{k-1} \neq 0 \\[4pt] \ldots \qquad\qquad \ldots \qquad\qquad \ldots \\[4pt] 1 + \lambda^{k-1} + \lambda^{k-1}\ \lambda^0 + \ldots + \lambda^{k-1}\ \lambda^0\ \ldots\ \lambda^{k-3} \neq 0 \end{array} \right.$$

with the following relations :

(LB) $\lambda^0 = \beta^0\ \dfrac{\beta^1 - 1}{\beta^0 - 1}$, $\lambda^1 = \beta^1\ \dfrac{\beta^2 - 1}{\beta^1 - 1}$, $\ldots$, $\lambda^{k-2} = \beta^{k-2}\ \dfrac{\beta^{k-1} - 1}{\beta^{k-2} - 1}$,

$$\lambda^{k-1} = \beta^{k-1}\ \frac{\beta^0 - 1}{\beta^{k-1} - 1}$$

(BL)

$$\left| \begin{array}{l} \beta^0 = (\lambda^0 + \lambda^0\ \lambda^1 + \ldots + \lambda^0\ \lambda^1\ \ldots\ \lambda^{k-1})/ \\[4pt] \qquad /(1 + \lambda^0 + \lambda^0\ \lambda^1 + \ldots + \lambda^0\ \lambda^1 \ldots\ \lambda^{k-2}) \\[6pt] \beta^1 = (\lambda^1 + \lambda^1\ \lambda^2 + \ldots + \lambda^1\ \lambda^2\ \ldots\ \lambda^0)/ \\[4pt] \qquad /(1 + \lambda^1 + \lambda^1\ \lambda^2 + \ldots + \lambda^1\ \lambda^2\ \ldots\ \lambda^{k-1}) \\[6pt] \ldots \quad \ldots \quad \ldots \quad \ldots \\[6pt] \beta^{k-1} = (\lambda^{k-1} + \lambda^{k-1}\ \lambda^0 + \ldots + \lambda^{k-1}\ \lambda^0\ \ldots\ \lambda^{k-2})/ \\[4pt] \qquad /(1 + \lambda^{k-1} + \lambda^{k-1}\ \lambda^0 + \ldots + \lambda^{k-1}\ \lambda^0\ \ldots\ \lambda^{k-3}) \ , \end{array} \right.$$

then :

$$\Longleftrightarrow \left| \begin{array}{l} \text{There exists } x \in \mathbb{C} \text{ such that for every } i \in \{0,1,\ldots,k-1\} \\[6pt] \left| \dfrac{x_{nk+i+1} - x}{x_{nk+i} - x} \text{ converges and } \beta^i = \lim_{n \to \infty} \dfrac{x_{nk+i+1} - x}{x_{nk+i} - x} \right| \\[10pt] \text{For every } i \in \{0,1,\ldots,k-1\} \ \dfrac{x_{nk+i+2} - x_{nk+i+1}}{x_{nk+i+1} - x_{nk+i}} \text{ converges} \\[6pt] \text{and} \\[4pt] \qquad\qquad \lambda^i = \lim_{n \to \infty} \dfrac{x_{nk+i+2} - x_{nk+i+1}}{x_{nk+i+1} - x_{nk+i}} \ . \end{array} \right.$$

Proposition 1'

If there exists $x \in C$ and $(\beta^0,\beta^1,\ldots,\beta^{k-1}) \in C^k$ satisfying (B) and such that, for every $i \in \{0,1,\ldots,k-1\}$:

$$\frac{x_{nk+i+1} - x}{x_{nk+i} - x} \text{ converges and } \lim_{n\to\infty} \frac{x_{nk+i+1} - x}{x_{nk+i} - x} = \beta^i \; ,$$

then, for every $i \in \{0,1,\ldots,k-1\}$:

$$\frac{x_{nk+i+2} - x_{nk+i+1}}{x_{nk+i+1} - x_{nk+i}} \text{ converges and } \lim_{n\to\infty} \frac{x_{nk+i+2} - x_{nk+i+1}}{x_{nk+i+1} - x_{nk+i}} = \lambda^i$$

with λ^i given by (LB) .

Proof

Let $i \in \{0,1,\ldots,k-2\}$

$$\frac{x_{nk+i+2} - x_{nk+i+1}}{x_{nk+i+1} - x_{nk+i}} =$$

$$= \frac{x_{nk+i+1} - x}{x_{nk+i} - x} \cdot \frac{\dfrac{x_{nk+i+2} - x}{x_{nk+i+1} - x} - 1}{\dfrac{x_{nk+i+1} - x}{x_{nk+i} - x} - 1} \underset{n\to\infty}{\to} \beta^i \, \frac{\beta^{i+1} - 1}{\beta^i - 1} = \lambda^i$$

The case $i = k-1$ is analogous.

Remark

The proposition is still true for sequences of a topological field.

Proposition 2'

If there exists $(\lambda^0,\lambda^1,\ldots,\lambda^{k-1}) \in C^k$ such that
$|\lambda^0 \, \lambda^1 \, \ldots \, \lambda^{k-1}| < 1$ and satisfying (L) , and if, for every $i \in \{0,1,\ldots,k-1\}$:

$$\frac{x_{nk+i+2} - x_{nk+i+1}}{x_{nk+i+1} - x_{nk+i}} \text{ converges to } \lambda^i \; ,$$

Then the sequence (x_n) converges to a limit x such that : for every $i \in \{0,1,..,k-1\}$,

$$\frac{x_{nk+i+1} - x}{x_{nk+i} - x} \text{ converges to } \beta^i \text{ , with } \beta^i \text{ given by (BL)}$$

Proof

We set $\lambda = \lambda^0 \lambda^1 \ldots \lambda^{k-1}$.

Let $n \in \mathbf{N}$.   From lemma 2 , the following series are absolutly convergent :

$$\sum_{p \in \mathbf{N}^*} \lambda_{nk} \lambda_{nk+1} \cdots \lambda_{nk+pk-1} \underset{n \to \infty}{\to} \lambda/(1-\lambda) \; ;$$

$$\sum_{p \in \mathbf{N}^*} \lambda_{nk+1} \lambda_{nk+2} \cdots \lambda_{nk+pk} \to \lambda/(1-\lambda) \; ;$$

$$\cdots \qquad \cdots \qquad \cdots$$

$$\sum_{p \in \mathbf{N}^*} \lambda_{nk+k-1} \lambda_{nk+k} \cdots \lambda_{nk+pk+k-2} \to \lambda/(1-\lambda) \; .$$

Thus, the series $\sum\limits_{p \in \mathbf{N}} \lambda_{nk} \lambda_{nk+1} \cdots \lambda_{nk+p} = \mu_{nk}$ is absoutly convergent.

$$\mu_{nk} = \lambda_{nk} + \lambda_{nk} \lambda_{nk+1} + \ldots + \lambda_{nk} \lambda_{nk+1} \cdots \lambda_{nk+k-2} +$$

$$\sum_{p \in \mathbf{N}^*} \lambda_{nk} \lambda_{nk+1} \cdots \lambda_{nk+pk-1} +$$

$$\lambda_{nk} \sum_{p \in \mathbf{N}^*} \lambda_{nk+1} \lambda_{nk+2} \cdots \lambda_{nk+pk} +$$

$$\cdots \qquad \cdots \qquad \cdots \qquad\qquad +$$

$$\lambda_{nk} \lambda_{nk+1} \cdots \lambda_{nk+k-2} \sum_{p \in \mathbf{N}^*} \lambda_{nk+k-1} \lambda_{nk+k} \cdots \lambda_{nk+pk+k-2}$$

Thus :

$$\lim_{n \to \infty} \mu_{nk} = \lambda^0 + \lambda^0 \lambda^1 + \ldots + \lambda^0 \lambda^1 \ldots \lambda^{k-2} +$$

$$\frac{\lambda}{1-\lambda} [1 + \lambda^0 + \lambda^0 \lambda^1 + \ldots + \lambda^0 \lambda^1 \ldots \lambda^{k-2}]$$

$$= \left[1 + \frac{\lambda}{1-\lambda}\right] (\lambda^0 + \lambda^0 \lambda^1 + \ldots + \lambda^0 \lambda^1 \ldots \lambda^{k-2}) + \frac{\lambda}{1-\lambda}$$

$$= \frac{\lambda^0 + \lambda^0 \lambda^1 + \ldots + \lambda^0 \lambda^1 \ldots \lambda^{k-1}}{1 - \lambda} \quad .$$

Since $(\mu_{nk})_{n\in\mathbf{N}}$ is convergent, from lemma 1 we obtain that (x_n) converges to x . Since (L) is satisfied, $\mu_{nk} \neq 1$, when n is large enough, hence (using lemma 1) :

$$\lim_{n\to\infty} \frac{x_{nk+1} - x}{x_{nk} - x} = \lim_{n\to\infty} \frac{\mu_{nk}}{1 + \mu_{nk}} = \frac{\lambda^0 + \lambda^0 \lambda^1 + \ldots + \lambda^0 \lambda^1 \ldots \lambda^{k-1}}{1 + \lambda^0 + \lambda^0 \lambda^1 + \ldots + \lambda^0 \lambda^1 \ldots \lambda^{k-2}} = \beta^0$$

The reasoning is similar for β^1 , β^2 , ..., β^{k-1} .

Remark

The proposition 2' is still true for sequences in a valued field.

Proposition 3'

If there exists $(\lambda^0, \lambda^1, \ldots, \lambda^{k-1}) \in \mathbf{C}^k$ such that

$$\left|\lambda^0 \lambda^1 \ldots \lambda^{k-1}\right| > 1$$

satisfying (L) and such that, for every $i \in \{0, 1, \ldots, k-1\}$:

$$\frac{x_{nk+i+2} - x_{nk+i+1}}{x_{nk+i+1} - x_{nk+i}} \quad \text{converges to } \lambda^i \ ,$$

then the sequence (x_n) is such that $\lim_{n\to\infty} |x_n| = +\infty$

and for every $x \in \mathbf{C}$, $i \in \{0, 1, \ldots, k-1\}$

$$\frac{x_{nk+i+1} - x}{x_{nk+i} - x} \quad \text{converges to } \beta^i \ ,$$

with β^i given by (BL).

Proof

We use the notation of lemma 1 and lemma 2 . We set :

$$\alpha^0 = \frac{1}{\lambda^0} \ , \ \ldots \ , \ \alpha^{k-1} = \frac{1}{\lambda^{k-1}} \ , \ \alpha = \frac{1}{\lambda^0} \frac{1}{\lambda^1} \ldots \frac{1}{\lambda^{k-1}} \ , \ \alpha_n = \frac{1}{\lambda_n}$$

We obtain :

$$\frac{x_{nk+1} - x_1}{x_{nk} - x_1} = \frac{\lambda_0 + \lambda_0 \lambda_1 + \ldots + \lambda_0 \lambda_1 \ldots \lambda_{nk-1}}{\lambda_0 + \lambda_0 \lambda_1 + \ldots + \lambda_0 \lambda_1 \ldots \lambda_{nk-2}}$$

$$= \frac{1 + \alpha_{nk-1} + \alpha_{nk-1} \alpha_{nk-2} + \ldots + \alpha_{nk-1} \alpha_{nk-2} \ldots \alpha_1}{\alpha_{nk-1} + \alpha_{nk-1} \alpha_{nk-2} + \ldots + \alpha_{nk-1} \alpha_{nk-2} \ldots \alpha_1}$$

As in the proof of the proposition 2' , and using lemma 2, we obtain that :

$$\lim_{n \to \infty} \frac{x_{nk+1} - x_1}{x_{nk} - x_1} =$$

$$= \frac{1 + \dfrac{\alpha}{1-\alpha} + \alpha^{k-1} \dfrac{1}{1-\alpha} + \alpha^{k-1} \alpha^{k-2} \dfrac{1}{1-\alpha} + \ldots + \alpha^{k-1} \ldots \alpha^1 \dfrac{1}{1-\alpha}}{\dfrac{\alpha}{1-\alpha} + \alpha^{k-1} \dfrac{1}{1-\alpha} + \alpha^{k-1} \alpha^{k-2} \dfrac{1}{1-\alpha} + \ldots + \alpha^{k-1} \ldots \alpha^1 \dfrac{1}{1-\alpha}}$$

$$= \frac{1 + \alpha^{k-1} + \alpha^{k-1} \alpha^{k-2} + \ldots + \alpha^{k-1} \alpha^{k-2} \ldots \alpha^1}{\alpha + \alpha^{k-1} + \alpha^{k-1} \alpha^{k-2} + \ldots + \alpha^{k-1} \alpha^{k-2} \ldots \alpha^1}$$

$$= \frac{\lambda^0 \lambda^1 \ldots \lambda^{k-1} + \lambda^0 \lambda^1 \ldots \lambda^{k-2} + \ldots + \lambda^0 \lambda^1 + \lambda^0}{1 + \lambda^0 \lambda^1 \ldots \lambda^{k-2} + \ldots + \lambda^0 \lambda^1 + \lambda^0} = \beta^0$$

and similar relations for $\beta^1, \beta^2, \ldots, \beta^{k-1}$.

We have :

$$\frac{x_{kn+1} - x_1}{(x_1 - x_0) \lambda_0 \lambda_1 \ldots \lambda_{kn-1}} = \frac{\lambda_0 + \lambda_0 \lambda_1 + \ldots + \lambda_0 \lambda_1 \ldots \lambda_{kn-1}}{\lambda_0 \lambda_1 \ldots \lambda_{kn-1}}$$

$$= 1 + \alpha_{nk-1} + \alpha_{nk-1} \alpha_{nk-2} + \ldots + \alpha_{nk-1} \ldots \alpha_1$$

Thus :

$$\lim_{n \to \infty} \frac{x_{kn+1} - x_1}{(x_1 - x_0)(\lambda_0 \lambda_1 \ldots \lambda_{kn-1})} =$$

$$= \frac{1}{1-\alpha} + \frac{\alpha^{k-1}}{1-\alpha} + \frac{\alpha^{k-1} \alpha^{k-2}}{1-\alpha} + \ldots + \frac{\alpha^{k-1} \alpha^{k-2} \ldots \alpha^1}{1-\alpha}$$

$$= \frac{1 + \alpha^{k-1} + \alpha^{k-1}\,\alpha^{k-2} + \dots + \alpha^{k-1}\,\alpha^{k-2}\dots\alpha^1}{1-\alpha}$$

$$= \frac{\lambda + \lambda^0\,\lambda^1 \dots \lambda^{k-2} + \lambda^0\,\lambda^1\dots\lambda^{k-3} + \dots + \lambda^0}{\lambda-1}$$

$$= \frac{\lambda^0(1 + \lambda^1 + \lambda^1\,\lambda^2 + \dots + \lambda^1\,\lambda^2\dots\lambda^{k-1})}{\lambda-1} \neq 0$$

Hence :
$$\lim_{n\to\infty} |x_{kn+1} - x_1| = +\infty \quad ;$$

and so :
$$\lim_{n\to\infty} |x_{kn+1}| = +\infty \ .$$

We proceed similarly for (x_{kn+2}) , (x_{kn+3}) ,..., (x_{kn+k}) .

hence :
$$\lim_{n\to\infty} |x_n| = +\infty \ .$$

The remainder of the proof is similar to the proof of proposition 3.

Remark

Proposition 3' is still true for sequences in a valued field.

(c) Applications

The general extrapolation algorithm due to T. HAVIE ([23]) and to C. BREZINSKI ([6]), and called the E-algorithm is defined by :

$$E_k(S_n) = \frac{\begin{vmatrix} S_n & S_{n+1} & \dots & S_{n+k} \\ g_1(n) & g_1(n+1) & \dots & g_1(n+k) \\ \dots & \dots & \dots & \dots \\ g_k(n) & g_k(n+1) & \dots & g_k(n+k) \end{vmatrix}}{\begin{vmatrix} 1 & 1 & \dots & 1 \\ g_1(n) & g_1(n+1) & \dots & g_1(n+k) \\ \dots & \dots & \dots & \dots \\ g_k(n) & g_k(n+1) & \dots & g_k(n+k) \end{vmatrix}}$$

where $(g_i(n))$ is a sequence of coefficients. Depending on $g_i(n)$,
we can obtain the Richardson transformation, or the G-transformation
([22]) ,   the Shanks transformation ([28]) , the Germain-Bonne
transformation ([21]) , the generalized Levin transformation ([26]) ,
the first generalization of the ε-algorithm ([4],[5]) , or the
p-process [4].   C. BREZINSKI proposed a recursive method for the
computation of $E_k(S_n)$ and gave the following theorem [6] :

> If (S_n) converges to S ; if, for every i , $g_i(n+1)/g_i(n)$
> converges to $b_i \neq 1$; if
>
> $\forall$ i,j   i $\neq$ j => $b_i \neq b_j$
>
> $\forall$ k $(E_{k-1}(S_{n+1}) - S)/(E_{k-1}(S_n) - S)$ converges to b_k ,
>
> Then : $E_k(S_n) - S = o (E_{k-1}(S_n) - S)$.

With theorem 1 , we now obtain :

> If (S_n) converges to S ; if, for every i $g_i(n+1)/g_i(n)$
> converges to $b_i \neq 1$; if
>
> $\forall$ i,j   i $\neq$ j => $b_i \neq b_j$
>
> $\forall$ k $(E_{k-1}(S_{n+1}) - E_{k-1}(S_n))/(E_{k-1}(S_n) - E_{k-1}(S_{n-1}))$
>
> converges to b_k ,
>
> Then : $E_k(S_n) - S = o (E_{k-1}(S_n) - S)$.

In the particular case of the Δ^2 of Aitken, we obtain :

> If $(S_{n+2} - S_{n+1})/(S_{n+1} - S_n)$ converges to λ , $|\lambda| < 1$,
>
> then $\varepsilon_2^{(n)}$ accelerates the convergence of (S_{n+1}) .

Practically, these results allow us to build programs for computing
acceleration arrays (as the ε-algorithm), in which the k-th column
is calculated only when the condition on the $(k-1)^{st}$ column is
satisfied.

Another application is given in section 2.

2 -   ACCELERATION OF PERIODICO-LINEAR SEQUENCES

The results of the first section lead us to define the family of
periodico-linear sequences, and to study the accelerability of this
family.

Definition

We say that the real sequence (x_n) is of periodico-linear type if there exists $k \in \mathbf{N}^*$ such that :

$$\exists\, x \in \mathbf{R}\ ,\ \exists\, (\beta^0, \beta^1, \ldots, \beta^{k-1}) \in \mathbf{R}^k\ ,\ \beta^0, \ldots, \beta^{k-1} \notin \{0, 1\}$$

$$\lim_{n \to \infty} \frac{x_{nk+1} - x}{x_{nk} - x} = \beta^0\ ;$$

$$\lim_{n \to \infty} \frac{x_{nk+2} - x}{x_{nk+1} - x} = \beta^1\ , \ldots,\ \lim_{n \to \infty} \frac{x_{nk+k} - x}{x_{nk+k-1} - x} = \beta^{k-1}\ .$$

The least integer k satisfying this relation is called the period of the sequence, the numbers $\beta^0, \beta^1, \ldots, \beta^{k-1}$ are called the ratios of the sequence.

Remarks

1°) To say that the sequence (x_n) is of periodico-linear type is equivalent to saying that the sequence $(x_{n+1} - x)/(x_n - x)$ is asymptotically periodic.

2°) If a sequence (x_n) is of periodico-linear type with period $k = 1$, then (x_n) is of linear type. If $|\beta^0| < 1$, (x_n) is convergent, hence it is a linearly convergent sequence. The notion of sequences of periodico-linear type is a generalization of the notion of linearly convergent sequences.

Proposition 4

Let (x_n) be a sequence of periodico-linear type with period k and ratios $\beta^0, \beta^1, \ldots, \beta^{k-1}$.

If $|\beta^0 \beta^1 \ldots \beta^{k-1}| < 1$, then (x_n) is convergent (we say that (x_n) is periodico-linearly convergent).

If $|\beta^0 \beta^1 \ldots \beta^{k-1}| > 1$, then $\lim_{n \to \infty} |x_n| = + \infty$

Proof

If $|\beta^0 \beta^1 \ldots \beta^{k-1}| > 1$, then, for every $i \in \{0, 1, \ldots, k-1\}$:

$$\lim_{n \to \infty} \left| \frac{x_{nk+k+i} - x}{x_{nk+i} - x} \right| = \beta^0 \beta^1 \ldots \beta^{k-1}\ ;$$

Thus : $\lim_{n\to\infty} |x_{nk+i} - x| = + \infty$ and so $\lim_{n\to\infty} |x_{nk+i}| = + \infty$;

Hence :
$$\lim_{n\to\infty} |x_n| = + \infty$$

Similarly,   if $|\beta^0 \beta^1 \ldots \beta^{k-1}| < 1$,   then,   for   every
$i \in \{0,1,\ldots,k-1\}$, the sequence $(x_{nk+i})_{n\in\mathbf{N}}$ is linearly convergent
with ratio $\beta = \beta^0 \beta^1 \ldots \beta^{k-1}$ and limit x ; thus (x_n) converges
to x .

Proposition 5

Let $k \in \mathbf{N}^*$. The sequence transformation defined by :

$$(\Delta^2_k) \qquad t_n = x_n - \frac{(x_n - x_{n-k})^2}{x_n - 2x_{n-k} + x_{n-2k}} \qquad n \geq 2k \ ,$$

accelerates the convergence of every periodico-linearly convergent
sequence, with period k , and gives the exact limit (when $n \geq 2k$,
$t_n = x$) for every sequence

$$x_n = x + a \prod_{i=0}^{n} \beta_i \ ,$$

when (β_i) is periodic with period k .

Proof

$$\frac{t_n - x}{x_n - x} = 1 + \left(\frac{x_{n-k} - x}{x_n - x} - 1\right) \frac{1}{1 - \dfrac{x_{n-k} - x_{n-2k}}{x_n - x_{n-k}}}$$

Thus $$\lim_{n\to\infty} \frac{t_n - x}{x_n - x} = 0$$

If $x_n = x + a \prod_{i=0}^{n} \beta_i$, β_i periodic with period k ,

then $$t_n = x + \left(a \prod_{i=0}^{n-2k} \beta_i\right) \beta^2 - \left(a \prod_{i=0}^{n-2k} \beta_i\right) \frac{(\beta^2 - \beta)^2}{\beta^2 - 2\beta+1} = x \ ,$$

where $\beta = \beta_i \beta_{i+1} \ldots \beta_{i+k-1}$ (β does not depend on i) .

Remark

The Δ^2_k process generalizes the Δ^2 of Aitken. However the Δ^2_k process is a particular case of the transformation

$$t_n = x_n - (x_{n+p+r} - x_n)/((x_{n+2p+r+q} - x_{n+p+r+q})/(x_{n+p+q} - x_{n+q}) -1),$$

where p,q,r are 3 parameters in $\mathbf{Z}$, $p+r \neq 0$.

When $p = 1$, $r = q = 0$, we obtain the Δ^2 of Aitken ; when $p = -k$, $r = q = 0$, we obtain the Δ^2_k. For every p,q,r, the sequence (t_n) accelerates (x_n), if (x_n) is a linearly convergent sequence (Another generalization of the Δ^2 of Aitken having similar properties as Δ^2_k, is given in $[31]$ (see also $[32]$)).

The Δ^2_k is difficult to use practically for k is not generally known. Now we propose transformations which we can use without knowing k.

They are obtained by a combination of the algorithm for the determination of the period and of the Δ^2_k.

Here we assume that the algorithms of chapter 2, § 4, have been normalized.

Algorithm Δ I3(ρ)

Step p (p $\geq$ 2)

The step p of the algorithm I3(ρ) is applied to the sequence :

$$y_n = \frac{x_n - x_{n-1}}{x_{n-1} - x_{n-2}} \qquad n \in \{2,3,\ldots,p\}$$

Let $k(p)$ be the result.

We set $t_p = x_p - (x_p - x_{p-k(p)})^2/(x_p - 2x_{p-k(p)} + x_{n-2k(p)})$

Notation

Let $\rho \in \mathbf{R}^{+*}$, $k \in \mathbf{N}$. We denote by $P_{\rho,k}$ the set of the sequences (x_n) which are periodico-linearly convergent, with period k and having ratios $\beta^0,\beta^1,\ldots,\beta^{k-1}$ such that :

$$\beta^0 \frac{\beta^1 - 1}{\beta^0 - 1} \quad , \quad \beta^1 \frac{\beta^2 - 1}{\beta^1 - 1} \quad , \quad \ldots \quad , \quad \beta^{k-1} \frac{\beta^0 - 1}{\beta^{k-1} - 1} \quad ,$$

are mutually distant, at least ρ apart.

When $k = 1$, $P_{\rho,k}$ is the set of linearly convergent sequences.

We set $P_\rho = \underset{k \in \mathbf{N}^*}{\cup} P_{\rho,k}$.

Obviously, $\rho < \rho' \Rightarrow (P_{\rho',k} \subset P_{\rho,k} \ , \ P_{\rho'} \subset P_\rho)$.

Theorem 3

i) For every sequence $(x_n) \in P_\rho$, the algorithm Δ I3(ρ) accelerates the convergence of (x_n) .

ii) For every sequence $(x_n) \in P_\rho$ having the following form :

$$x_n = x + a \prod_{i=0}^{n} \beta_i \ , \ \beta_i \text{ periodic,}$$

the sequence (t_n) obtained by Δ I3(ρ) satisfies

$$\dashv p_0 \in \mathbf{N} \ \forall \ p \geq p_0 : t_n = x .$$

Proof

i) If $(x_n) \in P_\rho$, there exists $p_0 \in \mathbf{N}$ such that, for every
   $p \geq p_0$, k is the period of (x_n) ; hence, from proposition 3 ,
   (t_n) accelerates the convergence of (x_n).

ii) Similar.

Remarks

1°) If the sequence (x_n) is linearly convergent, then the algorithm
Δ I3(ρ) (for every $\rho \in \mathbf{R}^{+*}$) is ultimately equivalent to the Δ^2 of
Aitken. Thus the Δ I3(ρ) is really interesting for periodico-linear
sequence with period $p \geq 2$.

2°) We can use the same idea from the algorithms I1, I2, I3, I4, J1,
J2,J'1, J'2 ([13] and chapter 2).

We obtain new acceleration algorithms, each of them having a specific
domain of efficiency. (I4 and J1 are of special interest).

3°) Since $\rho < \rho' \Rightarrow P_{\rho'} \subset P_\rho$, we are lead to choose ρ small.
But when ρ is small, p_0 is large. Hence, we have to choose ρ not
too small. For example ;

$$1/1000 \leq \rho \leq 1/10 .$$

$4°$) If $x_n = x + a_n \lambda^n$, $|\lambda| < 1$, (a_n) periodic, then the ε-algorithm accelerates (x_n) ([19]). But the ε-algorithm does not accelerate all periodico-linear sequences.

3 - OPTIMALITY OF THE Δ^2 OF AITKEN RELATIVELY TO THE ACCELERATION OF LINEARLY CONVERGENT SEQUENCES

In this section, all sequences are complex sequences. Instead of Lin(C) , we shall write Lin.

The aim of this section is to prove that the Δ^2 of Aitken is optimal (non-improvable) relative to the problem of the acceleration of linearly convergent sequences.

In (a), the Pennacchi and Germain-Bonne results show the algebraic optimality.

In (b), we give 3 propositions that prove that it is impossible to enlarge Lin into another accelerable family.

In (c) , we establish that for every $s > 0$, it is impossible to accelerate Lin with degree $s + 1$. Since the Δ^2 accelerates Lin with degree 1, the Δ^2 is still the optimal algorithm relative to Lin.

These results do not mean that the Δ^2 is the best algorithm of acceleration. They only mean that concerning the family Lin, it is impossible to find a simpler or a more efficient algorithm of acceleration. This does not contradict the existence of subfamilies of Lin accelerable with degree 2 ([20]), nor does it contradict the existence of algorithms less simple but equally efficient on Lin (columns of the θ-algorithm, [2],[3],[8]).

(a) Algebraical optimality

We recall some of the results of Pennacchi ([27]).

A transformation of type (p,m) is a transformation

$t_n = x_n + P(x_{n+1} - x_n , \ldots, x_{n+p} - x_{n+p-1})/Q(x_{n+1} - x_n , \ldots, x_{n+p} - x_{n+p-1})$,

where P and Q are homogeneous polynomials of degree m and m-1 respectively.

Pennacchi proves :

* There is no transformation of type (1,n) or (p,1) which accelerates Lin.

* The Δ^2 is the unique transformation of type (2,2) which accelerates Lin.

* Every transformation of type (2,m) , m ≥ 2 , which accelerates
Lin is equivalent to the Δ^2 (i.e. gives the same transformed
sequences).

Hence, the Δ^2 is the simplest rational transformation which accelerates Lin. The following result due to Germain-Bonne ([20])
confirms that the Δ^2 is algebraically the simplest transformation
accelerating Lin.

There is no transformation of the form :

$$t_n = x_n + g(x_{n+1} - x_n) \quad ,$$

(g continuous at 0) accelerating Lin.

(b) The impossibility of enlarging Lin

We recall the definition of Lin :

Lin = {(x) ∈ Conv(C) | ∄ λ, 0 < |λ| ≤ 1, λ ≠ 1

$$\text{and} \quad \lim_{n \to \infty}(x_{n+1}-x)/(x_n-x) = \lambda\}$$

Now we define :

Lin a = {(x_n) ∈ Conv(C) | ∄ λ, 0 ≤ |λ| ≤ 1, λ ≠ 1 ,

$$\lim_{n \to \infty}(x_{n+1}-x)/(x_n-x) = \lambda\}$$

Lin b = {(x_n) ∈ Conv(C) | ∄ λ, 0 < |λ| ≤ 1,

$$\lim_{n \to \infty}(x_{n+1}-x)/(x_n-x) = \lambda\},$$

and

Lin c = {(x_n) ∈ Conv(C) | ∄ λ, λ' : 0 < λ < λ' < 1,

$$\forall n \in \mathbf{N} : \quad \lambda < |x_{n+1}-x|/|x_n-x| < \lambda'\} \cup \text{Lin}$$

Proposition 6

There is no normal transformation which accelerates the convergence of
Lin a.

Proposition 7

There is no normal transformation which accelerates the convergence of
Lin b.

Proposition 8

There is no normal transformation which accelerates the convergence of Lin c.

Remarks

1°) Proposition 6 is only about the normalized Δ^2. It means that the (non-normalized) Δ^2 does not really accelerate the set of sequences such that

$$\lim_{n\to\infty}(x_{n+1}-x)/(x_n-x) = 0.$$

2°) We only prove that certain families containing Lin are not accelerable, but perhaps certain other "over families" of Lin are accelerable.

Proof

Propositions 6, 7, 8 follow immediatly from the results of chapter 5 §
4.

(c) Acceleration of degree 1+s on Lin

Since the Δ^2 accelerates Lin with the degree 1, it is natural to ask : "Is it possible to accelerate Lin with degree 2 , or 1+s for some s > 0 ?" The answer is NO. (For the definition of degree of accelerability, see page xxx).

Theorem 4

Let s > 0. There is no k-normal transformation which accelerates Lin with the degree 1+s.

Proof

We prove the result with k = 0. When k > 0 we use a technique similar to [10].

We assume that N is a normal transformation which accelerates Lin with the degree 1+s , s > 0.

Let (ℓ_n) be a strictly decreasing sequence of real numbers which converges to ℓ and such that :

$$0 < \ell < 1 \quad \text{and} \quad \forall\, n \in \mathbf{N} \quad 0 < \ell_n < 1.$$

Let (x^0_n) be a sequence defined by x^0_0 , x^0_1 , $x^0_1 > x^0_0$ such that :

$$\forall n \ (x^0_{n+2} - x^0_{n+1})/(x^0_{n+1} - x^0_n) = \ell_0 \ .$$

Its limit is :

$$x^0 = x^0_0 + (x^0_1 - x^0_0) \ (1 + \ell_0 + \ell^2_0 + \ldots + \ell^n_0 + \ldots) =$$
$$= x^0_0 + (x^0_1 - x^0_0)/(1 - \ell_0)$$

Let (t^0_n) be the transformed sequence of (x^0_n) by N.

For every integer m , we set :

$$y^0_m = x^0_{m-1} + (x^0_m - x^0_{m-1})/(1 - \ell_1)$$

We verify that :

$$y^0_m < x^0 = x^0_{m-1} + (x^0_m - x^0_{m-1})/(1 - \ell_0) \ .$$

There exists m_0 such that :

$$\left.\begin{array}{c} |t^0_{m_0} - y^0_{m_0}|/|x^0_{m_0} - y^0_{m_0}| \geq |x^0_{m_0} - y^0_{m_0}|^s \\ \\ \text{and} \\ \\ t^0_{m_0} > y^0_{m_0} \end{array}\right| \tag{0}$$

Indeed, for every integer m :

$$(x^0 - y^0_m)/(x^0 - x^0_m) = (\ell_0 - \ell_1)/(\ell_0 - \ell_1 \ell_0) = u > 0$$
$$|t^0_m - y^0_m|/|x^0_m - y^0_m| \geq |t^0_m - y^0_m|/|x^0_m - x^0|$$
$$\geq |x^0 - y^0_m + t^0_m - x^0|/|x^0_m - x^0|$$
$$\geq u - |t^0_m - x^0|/|x^0_m - x^0|$$
$$t^0_m - y^0_m = ((t^0_m - x^0)/(x^0 - x^0_m) + u)(x^0 - x^0_m)$$

We choose m_0 such that :

$$|t^0_{m_0} - x^0|/|x^0_{m_0} - x^0| \leq u/2 \quad \text{and} \quad |x^0_{m_0} - y^0_{m_0}|^s \leq u/2.$$

This is possible, since :

$$\lim_{n \to \infty}|t^0_m - x^0|/|x^0_m - x^0| = 0 \quad \text{and} \quad \lim_{n \to \infty}(x^0_m - y^0_m) = 0.$$

We define a sequence (x^1_n) by :

$$x^1{}_m = x^0{}_m \qquad\qquad \text{for every} \qquad m \leq m_0,$$

$$(x^1{}_{m+1} - x^1{}_m)/(x^1{}_m - x^1{}_{m-1}) = \ell_1 \quad \text{for every} \quad m \geq m_0 .$$

The sequence $(x^1{}_n)$ converges to the limit $x^1 = y^0{}_{m_0}$.

For every integer m , we set :

$$y^1{}_m = x^1{}_{m-1} + (x^1{}_m - x^1{}_{m-1})/(1 - \ell_2)$$

There exists $m_1 > m_0$ such that :

$$|t^1{}_{m_1} - y^1{}_{m_1}|/|x^1{}_{m_1} - y^1{}_{m_1}| \geq |x^1{}_{m_1} - y^1{}_{m_1}|^s$$

$$\text{and}$$

$$t^1{}_{m_1} > y^1{}_{m_1} \tag{1}$$

We define a sequence $(x^2{}_n)$ by :

$$x^2{}_m = x^1{}_m \qquad\qquad \text{for every} \qquad m \leq m_1$$

$$(x^2{}_{m+1} - x^2{}_m)/(x^2{}_m - x^2{}_{m-1}) = \ell_2 \quad \text{for every} \quad m \geq m_1 .$$

The sequence $(x^2{}_n)$ converges to the limit $x^2 = y^1{}_{m_1}$, etc...

We define :

$$(x_n) = (x^0{}_0, x^0{}_1, \ldots, x^0{}_{m_0}, x^1{}_{m_0+1}, \ldots, x^2{}_{m_1+1}, \ldots)$$

The sequence converges to $x = \lim_{n \to \infty} x^i$.

The sequence (x^i) is decreasing. The sequence (x_n) is linearly convergent, since :

$$\forall\, m \in \{m_i - 1, \ldots, m_{i+1} - 2\} : (x_{m+2} - x_{m+1})/(x_{m+1} - x_m) = \ell_{i+1}$$

The relations (0) , (1) , ..., prove that, for every i :

$$|t^i{}_{m_i} - y^i{}_{m_i}|/|x^i{}_{m_i} - y^i{}_{m_i}|^{1+s} \geq 1$$

Since $x^{i+1} = y^i{}_{m_i}$, we have :

$$|t^i{}_{m_i} - x^{i+1}|/|x^i{}_{m_i} - x^{i+1}|^{1+s} \geq 1 .$$

From $\quad x^i{}_{m_i} < x < x^{i+1} < t^i{}_{m_i}$,   we obtain

$$|t^i{}_{m_i} - x| / |x^i{}_{m_i} - x|^{1+s} \geq 1$$

Thus (x_n) is not accelerated by N with degree $1+s$.

Remark

The technique used in this proof is analogous to the technique used in chapters 2 and 3 ([9],[12]), and in chapter 5 ([17],[18]).

REFERENCES

[1]   AITKEN A.C.
"On Bernouilli's numerical solution of algebraic equations"
Proc. Roy. Soc. Edinburgh 46, 1926, pp. 289-305.

[2]   BREZINSKI C.
"Accélération des suites à convergence logarithmique"
C.R. Acad. Sc., Paris, a, 273, 1971, pp. 772-774.

[3]   BREZINSKI C.
"Conditions d'application et de convergence de procédés
d'extrapolation"
Numer. Math., 20, 1972, pp. 64-79.

[4]   BREZINSKI C.
"Accélération de la convergence en Analyse Numérique"
Lecture Notes in Mathematics, 584, Springer-Verlag,
Heidelberg, 1977.

[5]   BREZINSKI C.
"Algorithmes d'accélération de la convergence : étude
numérique"
Technip, Paris, 1978.

[6]   BREZINSKI C.
"A general extrapolation algorithm"
Numer. Math.,35, 1980, pp. 175-187.

[7]   BROMWICH T.J.
"An introduction to the theory of infinite series"
Mac Millan London, 1949, 2nd ed.

[8]   CORDELLIER F.
"Caractérisation des suites que la première étape du
θ-algorithme transforme en suites constantes"
C.R. Acad. Sc., Paris, 284, A, 1977, pp. 389-392.

[9]   DELAHAYE J.P.
"Quelques problèmes posés par les suites de points non
convergentes et algorithmes pour traiter de telles suites"
Thèse de 3e cycle, Lille, 1979.

[10]   DELAHAYE J.P.
"Accélération des suites telles que $\lambda \le \Delta x_{n+1} / \Delta x_n \le \mu$"
Publication ANO n° 17, Univ. des Sc. et Techn. de Lille, 1979.

[11]   DELAHAYE J.P.
"Liens entre la suite du rapport des erreurs et celle du
rapport des différences : démonstrations"
Publication ANO, n° 14, Univ. des Sc. et Techn. de Lille,
1979.

[12]   DELAHAYE J.P.
"Algorithmes pour suites non convergentes"
Numer. Math., 34, 1980, pp. 333-347.

[13]   DELAHAYE J.P.
"Détermination de la période d'une suite pseudo-périodique"
Bulletin de la Direction des Etudes et Recherches de l'E.D.F.,
C, n° 1, 1980, pp. 65-80.

[14]   DELAHAYE J.P.
"Liens entre la suite du rapport des erreurs et celle du
rapport des différences"
C.R. Acad. Sc. Paris, 290, 1980, pp. 343-346.

[15]   DELAHAYE J.P.
"Accélération des suites dont le rapport des erreurs est
borné"
Calcolo, 81, 1981, pp. 103-116.

[16]   DELAHAYE J.P.
"Optimalité du procédé Δ^2 d'Aitken pour l'accélération de
la convergence linéaire"
RAIRO Analyse Numérique, 15, 1981, pp. 321-330.

[17]   DELAHAYE J.P. et GERMAIN-BONNE B.
"Résultats négatifs en accélération de la convergence"
Numer. Math., 35, 1980, pp. 443-457.

[18]   DELAHAYE J.P. et GERMAIN-BONNE B.
"The set of logarithmically convergent sequences cannot
be accelerated"
SIAM Numerical Anal., 1982, pp. 840-844.

[19]   GENZ A.
"Application of the ε-algorithm to quadrature problems"
in "Padé approximants and their applications",
Graves-Morris ed., Academic Press, New York, 1973.

[20]   GERMAIN-BONNE B.
"Estimation de la limite de suites et formalisation des
procédés d'accélération de convergence"
Thèse, Lille, 1978.

[21]   GERMAIN-BONNE B.
"Transformations de suites"
RAIRO, R-1, 1973, pp. 84-91.

[22]   GRAY H.L. and ATCHISON T.A.
"The generalized G transform"
Math. Comp. 22, 1968, pp. 595-606.

[23]   HAVIE T.
"Generalized Neville Type extrapolation schemes"
BIT 19, 1979, pp. 204-213.

[24]   KOWALEWSKI C.
       "Accélération de la convergence pour certaines suites à
       convergence logarithmique"
       Lecture Notes in Mathematics 888, Springer—Verlag,
       Heidelberg, 1981, pp. 263—272.

[25]   KOWALEWSKI C.
       "Possibilités d'accélération de la convergence logarithmique"
       Thèse de 3e cycle, Lille, 1981.

[26]   LEVIN D.
       "Development of non—linear transformations for improving
       convergence of sequences"
       Intern. J. Comp. Math. B3, 1973, pp. 371—388.

[27]   PENNACCHI R.
       "Le transformazioni rationali di una successione"
       Calcolo, 5, 1968, pp. 37—50.

[28]   SHANKS D.I.
       "Non—linear transformations of divergent and slowly convergent
       sequences"
       J. Math. Phys. 36, 1955, pp. 1—42.

[29]   SMITH D.A. and FORD W.F.
       "Acceleration of linear and logarithmic convergence"
       SIAM J. Numer. Anal. 16, 1979, pp. 223—240.

[30]   WIMP J.
       "Sequence transformations and their applications"
       Academic Press, New York, 1981.

[31]   GRAY H.L. and CLARK W.D.
       "On a class of non linear transformations and their
       application to the evaluation of infinite series"
       Journal fo Research of the N.B.S., 73B, 1969, pp. 251—274.

[32]   JONES B.
       "A note on the T_{+m} transformation"
       Nonlinear Analysis, Theory, Methods and Ap. 6, 1982,
       pp. 303—305.

Chapter 7

**Automatic Selection
of Sequence Transformations**

INTRODUCTION

When faced with the great number of various methods for accelerating the convergence of sequences ([3],[4],[6],[18]) , and also with the problem of the choice of parameters for some of them (the Richardson process [8] and ρ-algorithm [2]) , the user is in a quite difficult position. If some precise information about the behavior of the sequence to be accelerated is known, it is possible to determine the most powerful acceleration method, but even in this case, several possibilities remain. The user can (if he has plenty of time !) try all the methods and choose the best one with the help of some test problems.

In practice, however, it very often happens that the choice is arbitrary and thus is not the best one for the problem.

In section 1, automatic processes for selecting a good method of acceleration among several sequence transformations are proposed. In section 2, we are interested specifically in the problem of the selection of a good sequence of parameters for the Richardson process.

Section 1 comes from [10], section 2 from [11].

1 - GENERAL METHODS

All the proposed processes work with the following scheme :

At the n-th step, the various transformations "in competition" $A_1, A_2, \ldots, A_k$, are applied to the sequence (S_n) of a metric space.

The transformed points $A^{(n)}{}_1, A^{(n)}{}_2, \ldots, A^{(n)}{}_k$ are obtained. Then one of the transformed points is selected by using various tests (several are defined and studied). This point must be the best one at this step.

The use of these automatic selection methods is more expensive than the use of a single transformation for accelerating the convergence because it is necessary to simultaneously use several transformations. However, this is not a drawback because acceleration transformations are generally not expensive to perform.

(a) General definitions for selection methods

Let $R^{(n)}{}_i$ be a relation (true or false depending on i and n) defined for all $i \in \{1, 2, \ldots, k\}$ and for all $n \in \mathbf{N}$, $n \geq n_0$.

Let us set the following count coefficients :

$$0_r(n)_i = \begin{cases} 0 & \text{if } n < n_0 \text{ or if } R^{(n)}_i \text{ is not true} \\ 1 & \text{if } n \geq n_0 \text{ and if } R^{(n)}_i \text{ is true.} \end{cases}$$

$$1_r(n)_i = \text{card}\{q \in \{0,1,\ldots,n\} \mid q \geq n_0 \text{ and } R^{(q)}_i \text{ is true}\}$$

$$2_r(n)_i = \begin{cases} 0 & \text{if } n < n_0 \text{ or if } R^{(n)}_i \text{ is not true} \\ \max\{q \in \{1,2,\ldots,n-n_0+1\} \mid R^{(n)}_i, R^{(n-1)}_i, \ldots, R^{(n-q+1)}_i \\ \text{are true }\} \\ \quad \text{if } n \geq n_0 \text{ and if } R^{(n)}_i \text{ is true.} \end{cases}$$

Note that even if $R^{(n)}_i$ is not defined, the count coefficients exist when $n \leq n_0$ and become zero. We have the following properties :

$$\forall\ i \in \{1,2,\ldots,k\}\ ,\ \forall\ n \in \mathbf{N}\ .$$

$$1_r(n)_i = \sum_{j=0}^{n} 0_r(j)_i\ ;\ 0 \leq 0_r(n)_i \leq 2_r(n)_i \leq 1_r(n)_i\ ;$$

$$n \geq n_0 \Rightarrow 1_r(n)_i \leq n-n_0+1.$$

It is possible to define other count coefficients. For example, a coefficient can be defined such that $R^{(n)}_i$ has more importance than $R^{(n-1)}_i$, $R^{(n-1)}_i$ has more importance than $R^{(n-2)}_i$, etc...

Let $A_1, A_2, \ldots, A_k$ be k sequence transformations. We denote by :

$$A^{(n)}_1(S_m)\ ,\ A^{(n)}_2(S_m), \ldots, A^{(n)}_k(S_m)$$

respectively, the sequence obtained by applying these transformations to the sequence (S_m). For convenience (see chapter 1), we shall also write $A^{(n)}_i$ instead of $A^{(n)}_i(S_m)$. All of the sequences in this chapter are sequences of a metric space (E,d).

If we substitute the relation $R^{(n)}_i$ for the relation :

$$(C^{(n)}_i)\ :\ A^{(n)}_i = A^{(n-1)}_i\ ,$$

we obtain other count coefficients denoted by $0_c(n)_i$, $1_c(n)_i$, $2_c(n)_i$.

Let $\ell \in \mathbf{N}$. If we substitute the relation $R^{(n)}_i$ for the relation :

$$(\ell D^{(n)}_i)\ :\ \max_{0 \leq q \leq \ell} d(A^{n-q}_i, A^{n-q-1}_i) = \min_{1 \leq j \leq k} \max_{0 \leq q \leq \ell} d(A^{n-q}_j, A^{n-q-1}_j)$$

(which is defined only if $n \geq \ell+1$), we obtain other count coefficients denoted by

$$O_\ell d(n)_i \ , \ ^1_\ell d(n)_i \ , \ ^2_\ell d(n)_i \ .$$

When $\ell = 0$, we denote by $D^{(n)}_i$ the relation $_\ell D^{(n)}_i$ and by $O_d(n)_i$, $^1_d(n)_i$, $^2_d(n)_i$, the associated count coefficients.

Let $f \in \{0,1,2\}$. With the help of the coefficients $^f_r(n)_i$, we define a new transformation A which selects one of the transformations A_i at each step.

<u>Step n</u> of the transformation $A = {}^f R(A_1, A_2, \ldots, A_k)$

We compute $^f_r(n)_1$, $^f_r(n)_2$, $\ldots$, $^f_r(n)_k$. Let $i(n)$ be the smallest i such that $^f_r(n)_i = \max_{1 \leq j \leq k} {}^f_r(n)_j$.

We choose $A^{(n)} = A^{(n)}_{i(n)}$.

When applied together with the above count coefficient and k sequence transformations, this scheme defined new sequence transformations :

$$O_C(A_1,A_2,\ldots,A_k) \ , \ {}^1C(A_1,A_2,\ldots,A_k) \ , \ {}^2C(A_1,A_2,\ldots,A_k)$$

$$O_\ell D(A_1,A_2,\ldots,A_k) \ , \ {}^1_\ell D(A_1,A_2,\ldots,A_k) \ , \ {}^2_\ell D(A_1,A_2,\ldots,A_k)$$

Remarks

1°) Sometimes it is possible to compute the count coefficients recursively. For example, the computation of $^1c(n)_i$ need

$$A^{(0)}_i \ , \ A^{(1)}_i \ , \ \ldots \ , \ A^{(n)}_i \ ;$$

but, in fact, it is possible to do it only with $^1c(n-1)_i$, $A^{(n)}_i$ and $A^{(n-1)}_i$, because of the obvious relations :

$$^1c(n)_i = {}^1c(n-1)_i \quad \text{if} \quad A^{(n)}_i \neq A^{(n-1)}_i$$

$$^1c(n)_i = {}^1c(n-1)_i + 1 \quad \text{if} \quad A^{(n)}_i = A^{(n-1)}_i$$

2°) In a selection method of sequence transformations, it is desirable to have a normalization, i.e. an eventual shift of the index such that the computation of $A^{(n)}_i$ uses the same points of the sequence (S_n) at each step. For example, points among $S_0, S_1, \ldots, S_n$ The calculation of $\varepsilon^{(n)}_2$ uses S_n, S_{n+1}, S_{n+2}, hence if ε_2 is one of the sequence transformations "in competition" we write

$$A^{(0)} = S_0 \ ; \ A^{(1)}_i = S_1 \ ;$$

$$A^{(n)}_i = \varepsilon^{(n-2)}_2 \ , \ \text{for} \ n \geq 2.$$

When this normalization is respected, the new obtained sequence
transformations are also normalized.

3°) It is possible, in practical cases, that one of the used transformations can no longer be applied (for example in the case of
division by zero). In such a case two kinds of strategies can be
employed.

When this is the case for the transformation A_j at the step n_0 :

(α) For all $n \geq n_0$, only the transformations A_i with
$i \neq j$ are considered.

(β) For all $n \geq n_0$, we take : $A^{(n)}{}_j = A^{(n_0)}{}_j$, $\mathbf{R}^{(n)}{}_j \Longleftrightarrow \mathbf{R}^{(n_0)}{}_j$.

4°) Methods of selection among a finite number of transformations are
the only ones studied here in. However, it is possible to study the
case of an infinite number of transformations.

5°) Other general methods of selection are presented in $[7]$.

Example 1

Let A_1, A_2 be two given sequence transformations.

$^0D(A_1, A_2)$ is the sequence transformation A which, at the step n ,
produces the more stable one of A_1 or A_2 ; i.e.

$$A^{(n)} = A^{(n)}{}_1 \quad \text{if} \quad d(A^{(n)}{}_1 , A^{(n-1)}{}_1) \leq d(A^{(n)}{}_2 , A^{(n-1)}{}_2)$$

$$A^{(n)} = A^{(n)}{}_2 \quad \text{if} \quad d(A^{(n)}{}_2 , A^{(n-1)}{}_2) < d(A^{(n)}{}_1 , A^{(n-1)}{}_1)$$

$^1C(A_1, A_2)$ is the sequence transformation A which, at the step n ,
produces the less frequently changed one of A_1 or A_2 ; i.e. :

$$A^{(n)} = A^{(n)}{}_1 \quad \text{if in the first } n \text{ steps } A^{(q)}{}_1 = A^{(q-1)}{}_1$$

is more frequently true than $A^{(q)}{}_2 = A^{(q-1)}{}_2$

$$A^{(n)} = A^{(n)}{}_2 \quad \text{if not.}$$

Example 2

In practice, the most interesting methods are the selection among
transformations as different as possible. We shall see a few computational experiments with

$$f_\ell D(A_1, A_2, A_3, A_4, A_5, A_6)$$

where :

A_1 is the Richardson process ([3],[4],[5],[14],[16])

A_2 is the ε-algorithm ([2],[4],[17],[18],[19])

A_3 is the ρ-algorithm ([2],[4],[5],[18])

A_4 is the Overholt process ([3],[9])

A_5 is the Θ-algorithm ([1],[3])

A_6 is the iterated Δ^2 ([4])

(b)   Two results on selection methods

We say that the sequence transformation   A   is **regular for the family**
S   of convergent sequences if :

$$\forall \ (S_n) \in S : \lim_{n \to \infty} S_n = \lim_{n \to \infty} A^{(n)}$$

We say that the sequence transformation   A   is **semiregular for the**
family S   of convergent sequences if :

$$\forall \ (S_n) \in S \ [\ \exists \ n_0 \in \mathbf{N}, \ \forall \ n \geq n_0 \ A^{(n)} = A^{(n+1)}]$$

$$\Rightarrow [\ \exists \ m_0 \in \mathbf{N} \ , \ \forall \ m \geq m_0 \ A^{(m)} = \lim_{n \to \infty} S_n] \ .$$

We say that the sequence transformation   A   is **exact for the family**   S
of convergent sequences if :

$$\forall \ (S_n) \in \mathbf{S} \ , \ \exists \ m_0 \in \mathbf{N} \ , \ \forall \ m \geq m_0 : A^{(n)} = \lim_{n \to \infty} S_n \ .$$

We have :

$$[A \ \text{exact}] \Rightarrow [A \ \text{regular}] \Rightarrow [A \ \text{semiregular}] \ .$$

Results concerning these properties are known for most of the acceleration methods ([3],[4],[9],etc).

Theorem 1

Let $S_1, S_2, \ldots, S_k$ be   k   families of convergent sequences. For each
i $\in$ {1,2,...,k} ,   let A_i be a sequence transformation, exact for
S_1 and semiregular for $S_1 \cup S_2 \cup \ldots \cup S_k$. Then the transformations $^1C(A_1, A_2, \ldots, A_k)$, $^2C(A_1, A_2, \ldots, A_k)$, $^1_1D(A_1, A_2, \ldots, A_k)$,
$^2_1D(A_1, A_2, \ldots, A_k)$ are exact for

$$S_1 \cup S_2 \cup \ldots \cup S_k.$$

Proof

We indicate a proof only for $^1C(A_1, A_2, \ldots, A_k)$, denoted by A. One can easily adapt this proof to the other cases.

Let $(x_n) \in S_1 \cup S_2 \cup \ldots \cup S_k$. Let I_0 be the set of integers i_0 such that A_i is exact for (x_n). From the assumptions, there is at least one element in I_0, but it is possible that there are several ones (because we have not supposed that the S_i are disjoint). Since I_0 is finite, there exists $p_0 \in \mathbf{N}$ such that :

$$\forall\, i_0 \in I_0\,,\ \forall\, p \geq p_0 : A^{(p)}_{i_0} = x \quad (x = \lim_{n \to \infty} x_m)\ .$$

Hence, for each $p \geq p_0$, $i_0 \in I_0$, we have :

(*) $$^1C^{(p)}_{i_0} \geq p - p_0\ .$$

Contrarily, if $i \notin I_0$, there exists an infinite set of integers p such that :

$$A^{(p)}_i \neq A^{(p-1)}_i$$

(this comes from the assumption that A is semiregular). Consequently, there exists an integer $p_1 \geq p_0$ such that, for each $i \notin I_0$, $p \geq p_1$,

(**) $$^1C^{(p)}_i < p - p_0\ .$$

From (*), (**) we obtain that, for every $p \geq p_1$, $i(p) \in I_0$, and then $A^{(p)} = x$.

Remarks

1°) Theorem 1 tells nothing about the selection methods ^0C and $^0{}_1C$, and it is possible, by counterexamples, to prove that the theorem is not generally true for ^0C and $^0{}_\ell D$.

2°) The integer p_1 in the proof cannot be determined without other assumptions; but in practice p_1 is close to the index p such that there exists :

$$A^{(p)}_i = S\ .$$

3°) The corollary of proposition 2 (chapter 5), shows that the hypothesis of semi-regularity cannot be omitted.

Now, we assume that all sequences (x_n) are convergent and satisfy :

$$\exists\, n_0 \in \mathbf{N}\,,\ \forall\, n \geq n_0 : x_0 \neq \lim_{n \to \infty} x_m \ \text{ and }\ x_n \neq x_{n+1}\ .$$

We say that the sequence transformation A accelerates the convergence of (x_n) (resp. Δ-accelerates) if :

$$\lim_{m \to \infty} d(A^{(n)}, \lim_{m \to \infty} x_m)/d(x_n, \lim_{m \to \infty} x_m) = 0$$

(respectively $\lim_{n \to \infty} d(A^{(n+1)}, A^{(n)})/d(x_{n+1}, x_n) = 0).$

When A accelerates (resp. Δ-accelerates) all the sequences (x_n) of
the family **S** , we say that A accelerates **S** (resp. Δ-accelerates
S). We say that the sequence transformation A is fair for **S** , if,
for every sequence $(x_n) \in$ **S** ,

   (F_1) either A accelerates and Δ-accelerates (x_n)

   (F_2) or $\exists\, \varepsilon > 0$, $\exists\, n_0 \in$ **N** , $\forall\, n \geq n_0$

       $d(A^{(n+1)}, A^{(n)})/d(x_{n+1}, x_n) \geq \varepsilon$.

When $E = $ **R** , a sufficient condition such that A be fair for **S** is :
all sequences in **S** are monotonic, and the sequence

       $(d(A^{(n+1)}, A^{(n)})/d(x_{n+1}, x_n))$

is always convergent to some limit ℓ. The reason is that either ℓ
$= 0$ and then A accelerates and Δ-accelerates (see $[3]$) or $\ell > 0$
and then (F_2) holds.

Theorem 2

Let $\ell \in$ **N**.

Let $A_1, A_2, \ldots, A_k$ be   k   sequence transformations. For each
$i \in \{1, 2, \ldots, k\}$, we assume that A_i accelerates and Δ-accelerates
S_i , a family of convergent sequences, and that A_i is fair for
$S_1 \cup S_2 \cup \ldots \cup S_k$. Then the transformations

   $^{0}_{\ell}D(A_1, A_2, \ldots, A_k)$, $^{1}_{\ell}D(A_1, A_2, \ldots, A_k)$, $^{2}_{\ell}D(A_1, A_2, \ldots, A_k)$

accelerate $S_1 \cup S_2 \cup \ldots \cup S_k$.

Lemma

Let (x_n) be a convergent sequence with limit x .

Let $(A^{(n)}_1)$, $(A^{(n)}_2)$,..., $(A^{(n)}_k)$ be k sequences which accelerate the convergence of (x_n).

Let $i : $ **N** $\to \{1, 2, \ldots, k\}$.

Then the sequence $(A^{(n)}_{i(n)})$ accelerates the convergence of (x_n).

Proof

Let $\varepsilon \in \mathbf{R}^{+*}$.

For each $i \in \{1,2,\ldots,k\}$, there is $m_i \in \mathbf{N}$ such that :

$$\forall\ m \geq m_i\ :\ d(A^{(m)}i,x)/d(x_m,x) \leq \varepsilon\ .$$

Letting $M = \max\{m_i \mid i \in \{1,2,\ldots,k\}\}$; we obtain :

$$\forall\ m \geq M\ :\ d(A^{(m)}_{i(m)},x)/d(x_m,x) \leq \varepsilon\ .$$

Remarks

$1°$) The lemma is not true with Δ-acceleration. Here is a counter-example :

$$x_n = 1/n,\ A^{(n)}_1 = 1/n^2\ ,\ A^{(n)}_2 = 1/2n^2\ ,\ i(n) = (3 + (-1)^n)/2.$$

We have :

$$\lim_{n\to\infty} d(A^{(n+1)}_1\ ,\ A^{(n)}_1)/d(x_{n+1},x_n) = 0\ ,$$

$$\lim_{n\to\infty} d(A^{(n+1)}_2\ ,\ A^{(n)}_2)/d(x_{n+1},x_n) = 0\ ,$$

$$\lim_{n\to\infty} d(A^{(n+1)}_{i(n+1)}\ ,\ A^{(n)}_{i(n)})/d(x_{n+1},x_n) = 1/2\ .$$

$2°$) It is not always possible to accelerate a union of accelerable families (see chapter 5).

Proof of the theorem 2

We indicate a proof only for $^0_\ell D(A_1,A_2,\ldots,A_k)$,   denoted by A . One can easily adapt this proof for the other cases.

Let $(x_n) \in S_1 \cup S_2 \cup \ldots \cup S_k$.

Let x be its limit.

Let I_0 be the set of integers $i \in \{1,2,\ldots,k\}$ such that A_i accelerates and Δ-accelerates (x_n) .

From the assumptions, I_0 is non empty. If $j \notin I_0$, there exists $\varepsilon_j \in \mathbf{R}^{+*}$ and $n_j \in \mathbf{N}$ such that :

$$\forall\ n \geq n_j\ ,\ d(A^{(n+1)}_j\ ,\ A^{(n)}_j)/d(x_{n+1},x_n) \geq \varepsilon_j.$$

We write $\varepsilon = \min\{\varepsilon_j \mid j \notin I_0\}$, $N = \max\{n_j \mid j \notin I_0.\}$

We obtain :

$$\forall \, n \geq N \, , \, \forall \, j \notin I_0 \, : \, d(A^{(n+1)}{}_j \, , \, A^{(n)}{}_j)/d(x_{n+1},x_n) \geq \varepsilon \, .$$

Similarly, there exists $N' \geq N$ such that :

$$\forall \, n \geq N' \, , \, \forall \, i \in I_0 \, : \, d(A^{(n+1)}{}_i \, , \, A^{(n)}{}_i)/d(x_{n+1},x_n) < \varepsilon \, .$$

This implies that, for each $n \geq N' + \ell$, $i(n) \in I_0$.

From the lemma, we can conclude that

$$\lim_{n \to \infty} d(A^{(n)}{}_{i(n)} \, , \, x)/d(x_n,x) = 0.$$

Remarks

1°) The assumptions of Theorem 2 are often satisfied in practice, but it is difficult to show, in general, that they are satisfied for specific transformations A_i and large S_i. Nevertheless, this theorem justifies and explains the efficiency of the selection methods ϱD , as we shall see in the practical cases of section 3.

2°) We can generalize the notion of fair transformations. We say that the sequence transformation A is h-fair $(h \in \mathbf{N})$ for S , if for every sequence $(x_n) \in S$,
$(_hF_1)$ either A accelerates and Δ-accelerates (x_n),
or $(_hF_2)$:

$$\dashv \, \varepsilon > 0 \, , \, \dashv \, n_0 \in \mathbf{N} \, \forall \, n \geq n_0$$

$$\max_{0 \leq r \leq h} \, d(A^{(n-r+1)}, A^{(n-r)})/dx_{n+1},x_n) \geq \varepsilon \, .$$

We obtain $[A \text{ fair}] <=> [A \text{ } 0\text{-fair}]$.

If $h \geq h'$, $[A \text{ } h'\text{-fair}] => [A \text{ } h\text{-fair}]$. The assumption "A h-fair" is more often true than the assumption "A fair". However, theorem 2 is still true for $\ell \geq h$.

(c) Numerical experiments

The computational experiences presented here were made with C. Brezinski's codes ($[4]$), and I would like to take this opportunity to thank him for his excellent advice and support.

Example 1

In table 1 we present the first 10 steps for the 6 transformations in competition (see section 1, example 3), when they are applied to the sequence

$$x_n = \frac{1}{2} \left(\frac{1}{2}\right)^2 + \frac{1}{4} \left(\frac{1.3}{2.4}\right)^2 + \ldots + \frac{1}{2n} \left(\frac{1.3 \ldots (2n-1)}{2.4 \ldots (2n)}\right)^2 \rightarrow .22005074\ldots$$

Table 1

S_n	$R(1/n)$	ε	$\rho(n)$	Ov.	θ	Δ^2 it.
.125	.125 1	.125 1	.125 1	.125 1	.125 1	.125 1
.16015625	.19531250 1	.16015625 2	.16015625 2	.16015625 2	.16015625 2	.16015625 2
.17643229	.21582031 2	.19046336 3	.22077047 1	.19046336 5	.17643229 6	.19046336 3
.18577830	.21959093 2	.19838255 5	X .22033281 (1)	.20227823 4	.21918327 3	.19838255 5
.19183451	.22002856 2	.20570155 6	X .22004511 (1)	.20782237 5	.21966574 3	.20936274 4
.19607526	.22005258 2	.20871905 6	X .22004902 (1)	.21095162 5	.21984779 3	.21158726 4
.19920946	X .22005113 (3)	.21158833 6	.22005079 2	.21293141 5	.22005077 1	.21590176 4
.20161980	.22005076 1	.21305538 6	.22005076 1	.21428050 5	X .22005078 (3)	.21659597 4
.20353087	.22005074 1	.21447121 6	X .22005074 (1)	.21524960 5	.22005077 3	.21839937 4
.20508314	.22005074 1	.21529441 6	X .22005074 (1)	.21597391 5	.22005078 3	.21861073 4

The selected transformation by the method ^0D at the step n^{th} step is indicated with an X . We see that this chosen transformation is always among the transformations R , ρ , or Θ , which are here the transformations which accelerate the convergence. Consequently, we can say that the choice is correct. The method ^1D is also correct; at the step n with $n \geq 3$, the chosen transformation is always the ρ-algorithm. The method ^2D gives exactly the same results that the method ^0D does.

Since we know that $x = .22005074$, we can determine, at each step, the exact rank of each transformation (this rank is indicated in Table 1 by a number under each transformed point). For example, at step 4 (the first significant one), the ρ-aalgorithm gives .22033281 and this is the best transformed point; the rank of the ρ-algorithm at step 4 is consequently 1. The method ^0D chooses the ρ-algorithm at step 4; this is the best possible choice. At steps 5 and 6, the choice is still the best possible, but at step 7, ^0D chooses the Richardson process, whose rank is 3. The rank sequence of chosen transformations is $(1,1,1,3,3,1,1)$. This is not the best possible rank sequence, which is $(1,1,1,1,1,1,1)$. However, it is a good rank sequence, because all ranks are ≤ 3 , and there are only three sequences transformations accelerating the convergence of (x_n). For the two following examples, we have only indicated the rank sequence of the chosen transformation when the method ^0D is applied.

Example 2

$$x_n = \exp(- \sqrt{n}/10\sqrt{2})/n.$$

Table 2

0	1	2	3	4	5	6	7	8	9	10	11	12	13	14	15	16	17	18	19	20	21	22	23	24
			2	1	1	3	3	1	1	1	1	1	2	2	3	3	3	3	3	3	3	3	2	2
25	26	27	28	29	30	31	32	33	34	35	36	37	38	39	40	41	42	43	44	45	46	47	48	49
3	3	2	3	3	1	1	1	4	2	3	3	2	1	3	1	2	2	2	2	1	2	2	2	2

Only three transformations accelerate (x_n) : ε-algorithm, ρ-algorithm, and Δ^2 iterated. When n is large enough, at each step n , ^0D chooses one of these three transformations.

Example 3

$$x_{2n} = (1/2n), \quad x_{2n+1} = (4n + 5)/(2n + 2)^2 .$$

Table 3

0	1	2	3	4	5	6	7	8	9	10	11	12	13	14	15	16	17	18	19	20	21	22	23	24
			5	2	2	2	3	3	4	4	3	3	3	3	3	3	3	3	4	3	2	3	2	3

25	26	27	28	29	30	31	32	33	34	35	36	37	38	39	40	41	42	43	44	45	46	47	48	49
2	2	2	2	2	2	2	2	2	2	2	2	2	2	2	2	2	2	2	2	2	2	2	2	2

Only two transformations accelerate (x_n) : ε-algorithm and ρ-algorithm. The best one (ε-alg) is not chosen because the transformed sequence is alternating; when n is large enough, the choice
at the step n is the ρ-algorithm.

2 – AUTOMATIC CHOICE OF SEQUENCES OF PARAMETERS IN THE RICHARDSON EXTRAPOLATION

The Richardson process ([3],[4],[14],[16]), which allows the polynomial extrapolation of a sequence, is used to accelerate the convergence of sequences (for example, the Romberg method). To use it, a
sequence of parameters (the points of interpolation) is chosen. Often,
there is no reason to choose any particular sequence.

Here we propose methods which make an automatic choice of the sequence
of parameters among a finite set of sequences of parameters.

Our main idea is the following : we consider exactness properties of
transformations defined by the Richardson process and then we define
our methods of choice in order to have as good as possible exactness
properties.

It is well known that the transformations which are exact on a big
family of sequences are accelerative on a larger family of sequences.
Consequently, the methods defined here may be used to accelerate the
convergence of sequences.
In the section (a), we study exactness properties of sequence transformations defined from the Richardson process.

In section (b), we propose a method for the selection between "k-th
columns" (k fixed) of the Richardson arrays corresponding to various
sequences of parameters.

In sections (c) and (d); we consider the selection between "diagonals"
and "fast diagonals".

(a) Exactness properties of the transformations obtained from the Richardson process

Here we study what are the transformations (in the sense of chapter 1 : from <u>one</u> sequence we obtain <u>one</u> sequence) that we can obtain from the Richardson array. In proposition 1, exactness properties of these transformation are given.

Henceforth, all the sequences are complex sequences.

Let (a_n) be a sequence of parameters :

$$\forall\, n,m \quad n \neq m \Rightarrow a_n \neq a_m \, ,$$

$$\lim_{n \to \infty} a_n = 0$$

For every sequence (x_n) , we denote by $P^{(n)}_k$ the interpolation polynomial of degree $\leq k$, defined by

$$P^{(n)}_k(a_n) = x_n \, , \ldots, \ P^{(n)}_k(a_{n+k}) = x_{n+k} \, .$$

The values of $P^{(n)}(x)$ can be computed with the formulas of Neville-Aitken :

$$P^{(n)}_0(x) = x_n$$

$$P^{(n)}_{k+1}(x) = \frac{(a_{n+k+1} - x)\, P^{(n)}_k(x) - (a_n - x)\, P^{(n+1)}_k(x)}{a_{n+k+1} - a_n}$$

We set $T^{(n)}_k = P^{(n)}_k(0)$ and obtain an array :

$$
\begin{array}{llll}
x_0 = T^{(0)}_0 & & & \\
 & T^{(0)}_1 & & \\
x_1 = T^{(1)}_0 & & T^{(0)}_2 & \\
 & T^{(1)}_1 & & T^{(0)}_3 \ \ldots \\
x_2 = T^{(2)}_0 & & T^{(1)}_2 & \\
 & T^{(2)}_1 & & T^{(1)}_3 \ \ldots \\
x_3 = T^{(3)}_0 & & T^{(2)}_2 & \\
\end{array}
$$

$$\ldots \quad \ldots \quad \ldots \quad \ldots$$

Let $\alpha(n)$, $\beta(n)$ be two sequences of integers. We denote by T^β_α the sequence transformation which computes $(T^{\beta(n)}_{\alpha(n)})$ from (x_n).

If $\alpha(n) = k$ (k fixed) and $\beta(n) = n$, we obtain the "k-th diagonal" transformation (denoted by $T^{(k)}$).
If $\alpha(n) = \beta(n) = n$, we obtain a transformation, which we call the "fast diagonal" transformation, and denote it by $T^{()}$.

We define the following families :

$$S_k = \{(x_n) \mid \exists\, \alpha_0, \alpha_1, \ldots, \alpha_k \quad \forall\, n \quad x_n = \sum_{j=0}^{k} \alpha_j \, a^j_n \}$$

$$S = \bigcup_{k \in \mathbf{N}} S_k$$

$$C_k = \{(x_n) \mid \exists\, \alpha_0, \alpha_1, \ldots, \alpha_k \quad \exists\, n_0 \quad \forall\, n \geq n_0 \quad x_n = \sum_{j=0}^{k} \alpha_j \, a^j_n \}$$

$$C = \bigcup_{k \in \mathbf{N}} C_k$$

Obviously, we have :

$$S_k \subset S_{k+1} \subset S$$

$$\cap \qquad \cap \qquad \cap$$

$$C_k \subset C_{k+1} \subset C$$

The next proposition generalizes theorem 23 of [3].

It comes from the fact that the transformations T^β_α are constructed using interpolation polynomials.

Proposition 1

1°) If $\liminf_{n \to \infty} \alpha(n) \geq k$ $(k \in \mathbf{N}$ fixed$)$

and if $\lim_{n \to \infty} \beta(n) = + \infty$,

then T^β_α is exact on C_k

(thus the "k-th column" transformation is exact on C_k)

2°) If $\lim_{n \to \infty} \alpha(n) = + \infty$, then

T^β_α is exact on S.

(thus the "k-th diagonal" transformation is exact on S)

3°) If $\lim_{n \to \infty} \alpha(n) = \lim_{n \to \infty} \beta(n) = + \infty$,

Then T^β_α is exact on C

(thus the "fast diagonal" transformation is exact on C.

Remark

We see that the most efficient transformation is the "fast diagonal" transformation. The reason is that this transformation interpolates at a large number of points and "forgets" the first points of the sequence (which perhaps are not exact). However, when we have a sequence which is regular even on the beginning, the most efficient transformation is the first diagonal transformation, which does not forget the past.

Proof

We only prove part 3° .

Let $(x_n) \in \mathbf{C}.$ There exist $k \in \mathbf{N}$, $n_0 \in \mathbf{N}$, such that :

$$\forall \, n \geq n_0 : x_n = \sum_{i=0}^{k} \alpha_i \, a^i_n = P(a_n) \ .$$

Let $n_1 \geq n_0$ be such that : $\forall \, n \geq n_1 : \alpha(n) \geq k$, $\beta(n) \geq n_0$.

If $n \geq n_0$, then $P(\beta(n))_{\alpha(n)}$ is a polynomial of degree $\leq \alpha(n)$ such that

$$P(\beta(n))_{\alpha(n)}(a_i) = x_i$$

for every $i \in \{\beta(n), \beta(n)+1, \ldots, \beta(n) + \alpha(n)\}$

Since all the (a_i) are distinct, we have $P(\beta(n))_{\alpha(n)} = P$,

thus $\qquad\qquad T(\beta(n))_{\alpha(n)} = \alpha_0 = \lim_{n \to \infty} x_m \ .$

(b) Selection among k-th columns

In this section, we define a method for choosing among k-th columns (k fixed), obtained from ℓ different sequences of parameters (ℓ fixed).

Let k, ℓ be the two fixed integers.

Let (^1a_n) , (^2a_n) , ..., $(^\ell a_n)$ be ℓ sequences of parameters such that :

$$\forall \, i \in \{1, 2, \ldots, \ell\} \ \lim_{n \to \infty} {}^i a_n = 0 \ , \ \forall \, m, n \ \ m \neq n \Rightarrow {}^i a_m \neq {}^i a_n \ .$$

For every sequence (x_n) , we consider the interpolation polynomials of degree $\leq k$,

$$^1P(n)_k \ , \ ^2P(n)_k \ , \ldots, \ ^\ell P(n)_k \ ,$$

defined by :

$$i_P(n)_k \, (i_{a_n}) = x_n \, , \ldots, \, i_P(n)_k(i_{a_{n+k}}) = x_{n+k} \, .$$

We can obtain these via the Neville-Aitken formulas.

We set $i_T(n)_k = i_P(n)_k(0)$. With the $i_T(n)_k$, we obtain ℓ arrays (as in section (a)).

For every $i \in \{1,2,\ldots,\ell\}$, we define :

$$i_{S_k} = \{(x_n) \mid \exists \, \alpha_0,\alpha_1,\ldots,\alpha_k \, , \, \forall \, n : x_n = \sum_{j=0}^{k} \alpha_j \, i_{a^k_n}\}$$

$$i_S = \bigcup_{k \in \mathbf{N}} i_{S_k}$$

$$i_{C_k} = \{(x_n) \mid \exists \, \alpha_0,\alpha_1,\ldots,\alpha_k \, , \, \exists \, n_0 \, \forall \, n \geq n_0 : x_n = \sum_{j=0}^{k} \alpha_j \, i_{a^k_n}\}$$

$$i_C = \bigcup_{k \in \mathbf{N}} i_{C_k}$$

Obviously, the k-th column transformation ^1T_k is exact on ^1C_k , ^2T_k is exact on $^2C_k,\ldots,^\ell T_k$ is exact on $^\ell C_k$.

Now we define a new transformation $A = S(^1T_k,^2T_k,\ldots,^\ell T_k)$, which is exact on $^1C_k \cup \ldots \cup {}^\ell C_k$.

Transformation $S(^1T_k,^2T_k,\ldots,^\ell T_k) = A$

Step n

* For every $i \in \{1,2,\ldots,\ell\}$, compute :

   $$i_c(n) = \text{card}\{j \in \{1,2,\ldots,n\} \mid \, i_P(j)_k(i_{a_{j+k+1}}) = x_{j+k+1}\}$$

* Let $i(n) \in \{1,2,\ldots\ell\}$ be such that :

   $$i(n)_c(n) = \max \{i_c(n) \mid i \in \{1,2,\ldots,\ell\}\}$$

* Let $A(n) = i(n)_{T_k}(n)$.

Remarks

1°) The idea of the method is the following :

At the step n , we determine the sequence of coefficients which in the previous step has most frequently found the (k+2)th points from (k+1) consecutive points, then we propose the point given by the k-th column of the array corresponding to this sequence of parameters.

2°)   The $i_c(n)$ can be computed by :

$$i_c(n) = i_c(n-1)+1 \quad \text{if} \quad {^i}P(n)_k({^i}a_{n+k+1}) = x_{n+k+1} \ ,$$

$$i_c(n) = i_c(n-1) \quad \text{if not .}$$

3°)   This selection algorithm is a particular case of the general selection algorithm of § 1.   The count coefficients are of type 1 (with type 0 no results can be obtained). In theorem 3, we do not use an hypothesis of mutual regularity as in theorem 1 or 2.   This is the advantage of the proposed method.

4°)   Various modifications and generalizations of this selection method are possible :

(α)   In the definition of $i_c(n)$, we can replace

$${^i}P(j)_k({^i}a_{j+k+1}) = x_{j+k+1} \ ,$$

by

$$\left| {^i}P(j)_k({^i}a_{j+k+1}) - x_{j+k+1} \right| = \min_{h \,\in\,\{1,2,\ldots,\ell\}} \left| {^h}P(j)_k({^h}a_{j+k+1}) - x_{j+k+1} \right|.$$

The obtained method is well adapted to acceleration problems.

(β)   Here the selection is based on interpolation at the point $^i a_{n+k+1}$.   It is easy to imagine other methods (and theorems analogous to theorem 3), based on interpolation at the point $^i a_{n+k+2}$ or the point $^i a_{n-1}$,   or at two points (for example, $^i a_{n+k+1}$ and $^i a_{n+k+2}$).

(γ)   With some modifications (progressive introduction of the transformations) it is possible to obtain a selection among an infinite set of transformations.

(δ)   Instead of polynomials, we can also use other interpolation functions.

Theorem 3

The transformation $S(^1T_k, {}^2T_k,\ldots, {}^\ell T_k)$ is exact on

$$^1C_k \cup {}^2C_k \cup \ldots \cup {}^\ell C_k.$$

Proof

Let $(x_n) \in {}^1C_k \cup {}^2C_k \cup \ldots \cup {}^\ell C_k.$

Let i_0 be such that $(x_n) \in {}^{i_0}C_k$, $x = \lim_{n\to\infty} x_n$

From proposition 1, there exists $n_0 \in \mathbf{N}$ such that, for every $n \geq n_0$ $^i o_T(n) = x$ and $^i o_P(n)_k = {}^i o_{P_k}$, where $^i o_{P_k}$ is the polynomial such that for n large enough :

$$x_n = {}^i o_{P_k}({}^i o_{a_n}),$$

For $n \geq n_0$, we have $^i o_C(n) \geq n - n_0$.

Let I be the set of integers $i \in \{1, 2, \ldots, \ell\}$ such that there exists m_i satisfying :

$$\forall\, n \geq m_i \quad {}^i p(n)_k \,({}^i a_{n+k+1}) = x_{n+k+1}\ .$$

Let $j \notin I$.

Then, there is an infinite set of integers n such that :

$$j p(n)_k({}^i a_{n+k+1}) \neq x_{n+k+1}\ .$$

Hence, there exists $p_j \geq n_0$ such that, for every $n \geq p_j$:

$$j_c(n) \leq n - n_0.$$

Since this is true for every $j \notin I$, there exists

$$n_1 = \max\{p_j \mid j \in I\}$$

such that, for every $n \geq n_1$:

$$(*) \qquad\qquad\qquad i(n) \in I\ .$$

Now, let $i \in I$.

For every $n \geq m_i$:

$${}^i p(n)_k({}^i a_n) = x_n\ ,\ \ {}^i p(n)_k({}^i a_{n+1}) = x_{n+1}, \ldots, {}^i p(n)_k({}^i a_{n+k+1}) = x_{n+k+1}\ ,$$

$${}^i p(n+1)_k({}^i a_{n+1}) = x_{n+1}\ ,\ \ {}^i p(n+1)_k({}^i a_{n+2}) = x_{n+2}, \ldots, {}^i p(n+1)_k({}^i a_{n+k+2})$$

$$= x_{n+k+2}\ .$$

The polynomials $^i p(n+1)_k$, $^i p(n)_k$ match on $(k+1)$ points, thus they are equal to a polynomial $^i P_k$ which does not depend on n . For $n \geq m_i$, we have :

$$x_n = {}^i P_k({}^i a_n)\ .$$

When $n \to \infty$, we obtain

$$x = {}^i P_k(0).$$

With (*), we obtain that, for every $n \geq n_1$, $n \geq \max\{m_i \mid i \in I\}$,

$$^{i(n)}T(n)_k = {}^{i(n)}P(n)_k(0) = {}^{i(n)}P_k(0) = x.$$

(c) Selection among k-th diagonal transformations

A method analogous to the one of section (b) is used in order to select k-th diagonal transformations (k fixed) obtained from ℓ sequences of parameters (ℓ fixed).

Let k, ℓ be two fixed integers.

Let (^1a_n) , $(^2a_n), \ldots, (^\ell a_n)$ be ℓ sequences of parameters as in section (b).

For every given sequence (x_n) , we consider the polynomials of interpolation of degree $\leq n$,

$$^1P(k)_n , \ {}^2P(k)_n, \ldots, {}^\ell P(k)_n ,$$

defined by :

$$^iP(k)_n(^ia_k) = x_k , \ldots, \ {}^iP(k)_n(^ia_{n+k}) = x_{n+k} .$$

We obtain ℓ k-th diagonal transformations denoted (accordingly to section (a)), by $^1T(k), \ldots, {}^\ell T(k)$.

From proposition 1, the transformation $^1T(k)$ is exact on ^1S , ^2T_k on $^2S, \ldots, {}^\ell T(k)$ on $^\ell S$.

The new transformation $A = S(^1T(k), {}^2T(k), \ldots, {}^\ell T(k))$ will be exact on $^1S \cup {}^2S \cup \ldots \cup {}^\ell S$.

Transformation $S(^1T(k) , \ {}^2T(k) , \ldots, \ {}^\ell T(k)) = A.$

Step n

* For every $i \in \{1, 2, \ldots, \ell\}$, compute :

$$^ic(n) = \mathrm{card}\{j \in \{1, 2, \ldots, n\} \mid {}^iP(k)_j(^ia_{j+k+1}) = x_{j+k+1}\}$$

* Let $i(n) \in \{1, 2, \ldots, \ell\}$ be such that :

$$^{i(n)}c(n) = \max\{{}^ic(n) \mid i \in \{1, 2, \ldots, \ell\}\} .$$

* Let $A^{(n)} = {}^{i(n)}T(k)_n$.

Remarks analogous to those of section (b) may be made.

Theorem 4

The transformation $S(^1T(k)$, $^2T(k)$, ... , $^\ell T(k))$ is exact on

$$^1S \cup {}^2S \cup \ldots \cup {}^\ell S.$$

Proof

Let $(x_n) \in {}^1S \cup {}^2S \cup \ldots \cup {}^\ell S$.

Let i_0 be such that $(x_n) \in {}^{i_0}S$; $x = \lim_{n \to \infty} x_n$.

From proposition 1, there exists $n_0 \in \mathbf{N}$ such that , for every $n \geq n_0$,

$$^{i_0}T(k)_n = x \quad \text{and} \quad {}^{i_0}P(k)_n = {}^{i_0}P(k) ,$$

where $^{i_0}P_k$ is the polynomial such that :

$$x_n = {}^{i_0}P(k)({}^{i_0}a_n)$$

for n sufficiently large.

For $n \geq n_0$, we have $^{i_0}C(n) \geq n - n_0$.

As in theorem 3, we prove that there exists n_1 such that

$$n \geq n_1 \Rightarrow i(n) \in I ,$$

where I is the set of integers i such that there exists m_i :

$$\forall m \geq m_i : {}^iP(k)_n({}^ia_{n+k+1}) = x_{n+k+1} .$$

Let $i \in I$.

For every $n \geq m_i$:

$$^iP(k)_n({}^ia_k) = x_k , \quad {}^iP(k)_n({}^ia_{k+1}) = x_{k+1}, \ldots, {}^iP(k)_n({}^ia_{n+k+1}) = x_{n+k+1}$$

$$^iP(k)_{n+1}({}^ia_k) = x_k , \quad {}^iP(k)_{n+1}({}^ia_{k+1}) =$$

$$= x_{k+1}, \ldots, {}^iP(k)_{n+1}({}^ia_{n+k+2}) = x_{n+k+2}$$

The polynomial $^iP(k)_{n+1}$ (of degree $\leq n+1$) matches the polynomial $^iP(k)_n$ (of degree $\leq n$) on $n+2$ points; thus these two polynomials are equal to a polynomial $^iP(n)$ which does not depend on n .

We conclude as in theorem 3.

(d) Selection among fast diagonal transformations

Using a selection test with two points, we define a selection method among fast diagonal transformations.

Let ℓ be a fixed integer.

Let $(^1a_n),(^2a_n),\ldots,(^\ell a_n)$ be ℓ sequences of parameters as before.

For every given sequence (x_n) , we consider the polynomials of interpolation of degree $\leq n$, $^1P(n)_n, ^2P(n)_n,\ldots, ^\ell P(n)_n$, defined by :

$$^iP(n)_n(^ia_n) = x_n,\ldots, ^iP(n)_n(^ia_{2n}) = x_{2n}.$$

We obtain ℓ fast diagonal transformations, denoted (accordingly to section 1), by

$$^1T(),\ldots, ^\ell T() .$$

From proposition 1, the transformation $^1T()$ is exact on ^1C , $^2T()$ on $^2C,\ldots, ^\ell T()$ on $^\ell C$.

The new transformation

$$A = S^1(^1T(), ^2T(),\ldots, ^\ell T())$$

will be exact on :

$$^1C \cup {}^2C \cup \ldots \cup {}^\ell C .$$

Transformation $S^1(^1T(), ^2T(),\ldots, ^\ell T()) = A).$

Step n

* For every $i \in \{1,2,\ldots,\ell\}$, compute

$$^ic(n) = \text{card}\{j \in \{1,2,\ldots,n\}|^iP(j)_j(^ia_{2j+1}) = x_{2j+1}$$
$$\text{and } {}^iP(j)_j(^ia_{2j+2}) = x_{2j+2})\}$$

* Let $i \in \{1,2,\ldots,\ell\}$ be such that :

$$i(n)_c(n) = \max\{^ic(n)|i \in \{1,2,\ldots,\ell\}\} .$$

* Let $A^{(n)} = {}^{i(n)}T(n)_n$.

Theorem 5

The transformation $S^1(^1T(), ^2T(),\ldots, ^\ell T())$ is exact on

$$^1C \cup {}^2C \cup \ldots \cup {}^\ell C .$$

Proof

Let $(x_n) \in {}^1C \cup {}^2C \cup \ldots \cup {}^{\ell}C.$

Let i_0 be such that $(x_n) \in {}^{i_0}C$; $x = \lim\limits_{n \to \infty} x_n.$

From proposition 1, there exists $n_0 \in \mathbf{N}$ such that, for every $n \geq n_0$, ${}^{i_0}T(n) = x$ and ${}^{i_0}P(n)_n = {}^{i_0}P$, where ${}^{i_0}P$ is the polynomial such that

$$x_n = {}^{i_0}P({}^{i_0}a_n) \quad \text{for } n \text{ sufficiently large.}$$

For $n \geq n_0$, we have ${}^{i_0}c(n) \geq n - n_0.$

As in theorem 3, we prove that there exists n_1 such that :

$$n \geq n_1 \Rightarrow i(n) \in I ,$$

where I is the set of integers i such that there exists m_i for which :

$$\forall\, n \geq m_i \quad {}^iP(n)_n({}^ia_{2n+1}) = x_{2n+1} , \quad {}^iP(n)_n({}^ia_{2n+2}) = x_{2n+2}$$

Let $i \in I$. For every $n \geq m_i$:

$${}^iP(n)_n({}^ia_n) = x_n , \quad {}^iP(n)_n({}^ia_{n+1}) = x_{n+1}, \ldots, {}^iP(n)_n({}^ia_{2n+2}) = x_{2n+2}$$

$${}^iP(n+1)_{n+1}({}^ia_{n+1}) = x_{n+1}$$

$${}^iP(n+1)_{n+1}({}^ia_{n+2}) = x_{n+2}, \ldots, {}^iP(n+1)_{n+1}({}^ia_{2n+4}) = x_{2n+4}$$

The polynomial ${}^iP(n+1)_{n+1}$ (of degree $\leq n+1$) matches the polynomial ${}^iP(n)_n$ (of degree $\leq n$) on $n+2$ points; thus these two polynomials are equal to a polynomial ${}^iP$ which does not depend on n .

We conclude as in theorem 3.

REFERENCES

[1]   BREZINSKI C.
      "Accélération de suites à convergence logarithmique"
      C.R. Acad. Sc. Paris, A 273, 1971, pp.727-730.

[2]   BREZINSKI C.
      "Etudes sur les ε et ρ-algorithmes"
      Numer. Math., 17, 1971, pp. 153-162.

[3]   BREZINSKI C.
      "Accélération de la convergence en analyse numérique"
      Lecture Notes in Mathematics, 582, Springer-Verlag,
      Heidelberg, 1977.

[4]   BREZINSKI C.
      "Algorithmes d'accélération de la convergence.
      Etude numérique"
      Technip, Paris, 1978.

[5]   BREZINSKI C.
      "Analyse Numérique discrète"
      Cours polycopié, Lille, 1978.

[6]   BREZINSKI C.
      "A general extrapolation algorithm"
      Numer. Math. 35, 1980, pp. 175-187.

[7]   BREZINSKI C.
      "Error control in convergence acceleration processes"
      I.M.A. J. Num. Anal., to appear.

[8]   CORDELLIER F.
      "Caractérisation des suites que la première étape
      θ-algorithme transforme en suites constantes"
      C.R. Acad. Sc. Paris, t 284, 1977, pp. 389-392.

[9]   CORDELLIER F.
      "Sur la régularité des procédés δ^2 d'Aitken et
      W. de Lubkin"
      Padé Approximation and its applications.
      Lecture Notes in Mathematics, 765, Springer-Verlag,
      Heidelberg, 1980, pp. 20-35.

[10] DELAHAYE J.P.
      "Automatic selection of sequence transformations"
      Math. of Computation 37, 1981, pp. 197-204.

[11]   DELAHAYE J.P.
        "Choix automatique entre suites de paramètres dans
        l'extrapolation de Richardson"
        Padé Approximations and its applications,
        Lecture Notes in Mathematics, 888, Springer-Verlag,
        Heidelberg, 1981, pp. 158-172.

[12]   DELAHAYE J.P.
        "Algorithmes pour suites non convergentes"
        Numer. Math. 34, 1980, pp. 333-347.

[13]   DELAHAYE J.P. and GERMAIN-BONNE B.
        "Résultats négatifs en accélération de la convergence"
        Numer. Math. 35, 1980, pp. 443-457.

[14]   LAURENT J.P.
        "Etudes des procédés d'extrapolation en analyse numérique"
        Thèse, Grenoble, 1964.

[15]   OVERHOLT K.J.
        "Extended Aitken Acceleration"
        B.I.T. V. 6, 1965, pp. 122-132.

[16]   RICHARDSON L.F.
        "The deferred approach to the limit"
        Trans. Phil. Roy. Soc. 226, 1927, pp. 261-299.

[17]   SHANKS D.
        "Non linear transformations of divergent and slowly
        convergent sequences"
        J. Math. Phys. 34, 1955, pp. 1-42.

[18]   WIMP J.
        "Sequence transformations and their applications"
        Academic Press, New York, 1981.

[19]   WYNN P.
        "On a device for computing the $e_m(s_n)$ transformation"
        MTAC 16, 1956, pp. 91-96.

INDEX

Springer Series in Computational Mathematics

Editorial Board: **R. L. Graham, J. Stoer, R. Varga**

Computational Mathematics is a series of outstanding books and monographs which study the applications of computing in numerical analysis, optimization, control theory, combinatorics, applied function theory, and applied functional analysis. The connecting link among these various disciplines will be the use of high-speed computers as a powerful tool. The following list of topics best describes the aims of *Computational mathematics:* finite element methods, multigrade methods, partial differential equations, multivariate splines and applications, numerical solutions of ordinary differential equations, numerical methods of optimal control, nonlinear programming, simulation techniques, software packages for quadrature, and p.d.e. solvers.

Computational Mathematics is directed towards mathematicians and appliers of mathematical techniques in disciplines such as engineering, computer science, economics, operations research and physics.

Volume 1

R. Piessens, E. de Doncker-Kapenga, C. W. Überhuber, D. K. Kahaner

QUADPACK

A Subroutine Package for Automatic Integration

1983. 26 figures. VII, 301 pages. ISBN 3-540-12553-1

Contents: Introduction. – Theoretical Background. – Algorithm Descriptions. – Guidelines for the Use of QUADPACK. – Special Applications of QUADPACK. – Implementation Notes and Routine Listings. – References.

Volume 2

J. R. Rice, R. F. Boisvert

Solving Elliptic Problems Using ELLPACK

1985. 53 figures. X, 497 pages. ISBN 3-540-90910-9

Contents: The ELLPACK System. – The ELLPACK Modules. – Performance Evaluation. – Contributor's Guide. – System Programming Guide. – Appendices. – Index.

Springer-Verlag
Berlin Heidelberg New York
London Paris Tokyo

Volume 3

N. Z. Shor

Minimization Methods for Non-Differentiable Functions

Translated from the Russian by K. C. Kiwiel, A. Ruszczyński

1985. VIII, 162 pages. ISBN 3-540-12763-1

Contents: Introduction. – Special Classes of Nondifferentiable Functions and Generalizations of the Concept of the Gradient. – The Subgradient Method. – Gradient-type Methods with Space Dilation. – Applications of Methods for Nonsmooth Optimization to the Solution of Mathematical Programming Problems. – Concluding Remarks. – References. – Subject Index.

Volume 4

W. Hackbusch

Multi-Grid Methods and Applications

1985. 43 figures, 48 tables. XIV, 377 pages. ISBN 3-540-12761-5

Contents: Preliminaries. – Introductory Model Problem. – General Two-Grid Method. – General Multi-Grid Iteration. – Nested Iteration Technique. – Convergence of the Two-Grid Iteration. – Convergence of the Multi-Grid Iteration. – Fourier Analysis. – Nonlinear Multi-Grid Methods. – Singular Perturbation Problems. – Elliptic Systems. – Eigenvalue Problems and Singular Equations. – Continuation Techniques. – Extrapolation and Defect Correction Techniques. – Local Techniques. – The Multi-Grid Method of the Second Kind. – Bibliography. – Subject Index.

Volume 5

V. Girault, P-A. Raviart

Finite Element Methods for Navier-Stokes Equations

Theory and Algorithms

1986. 21 figures. X, 374 pages. ISBN 3-540-15796-4

Contents: Mathematical Foundation of the Stokes Problem. – Numerical Solution of the Stokes Problem in the Primitive Variables. – Incompressible Mixed Finite Element Methods for Solving the Stokes Problem. – Theory and Approximation of the Navier-Stokes Problem. – References. – Index of Mathematical Symbols. – Subject Index.

Springer

Springer-Verlag
Berlin Heidelberg New York
London Paris Tokyo

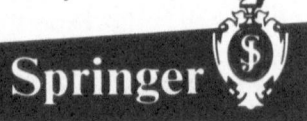

Springer